Marcos Glauser

Angaité's responses to deforestation

The Gran Chaco, the second largest biome of South America, entered a phase of deep and fast environmental changes a few decades ago. Indigenous peoples are amongst those most affected. This dissertation focuses on the responses of the Angaité of La Patria to altered access, use and management of natural resources inside and outside their colony over the past 20 years (1995 – 2015). From a third-generation political ecologists' perspective, I consider the Angaité's adaptation a transformation of cosmographical practices because the latter contribute to the production of a particular place or territory and a particular understanding of the world. I focus mainly on two types of cosmographical practices: ritual practices, like the choqueo, and livelihood related practices. I explore how the Paraguayan state allowed and promoted, by de facto not recognising "La Patria territory", resource use practices that were in-compatible with certain Angaité livelihood practices. I claim that, by redefining the "political" and the "social collective", the adaptation of livelihood and land use strategies and the new patronage networks with the state and NGOs impacted the Angaité cosmography, forms of leadership, and networks of sharing. This redefinition manifests in everyday contradictions, tensions and internal political conflicts. Finally, I argue that the responses from the Angaité should be understood as part of a longer and wider process of adaptive resistance. This paradoxical term refers to the ways of sustaining existence in interaction with other worlds. The Angaité, aware of and concerned by the disempowerment these changes provoke, are trying to regain protagonism.

CURUPIRA

herausgegeben von

Curupira –
Förderverein Kultur-
und Sozialanthropologie
in Marburg e.V.

Band 31

LIT

Marcos Glauser

Angaité's responses to deforestation

Political ecology of the livelihood and land use strategies of an indigenous community from the Paraguayan Chaco

LIT

This research was supported, printed and/or published by the BMZ via DAAD, the Carl and Charlotte Schott Stiftung of the Philipps-Universität Marburg and a grant from the Dr. Hermann Eiselen Doctoral Programm of the Foundation fiat panis.

Dissertation im Fachbereich Gesellschaftswissenschaften und Philosophie der Philipps-Universität Marburg

Curupira – Förderverein „Kultur- und Sozialanthropologie in Marburg“ e.V.
c/o Institut für Vergleichende Kulturforschung
Fachgebiet Kultur- und Sozialanthropologie
Deutschhausstraße 3, 35037 Marburg/Lahn
Tel. 06421/282-22082, Fax: 06421/282-2140
E-Mail: info@curupira.de
www.curupira.de

Bibliographic information published by the Deutsche Nationalbibliothek
The Deutsche Nationalbibliothek lists this publication in the Deutsche Nationalbibliografie; detailed bibliographic data are available on the Internet at http://dnb.d-nb.de.

ISBN 978-3-643-91011-0 (pb)
ISBN 978-3-643-96011-5 (PDF)
Zugl.: Marburg, Univ., Diss., 2018

A catalogue record for this book is available from the British Library.

Klosbachstr. 107
CH-8032 Zürich
Tel. +41 (0) 44-251 75 05
E-Mail: zuerich@lit-verlag.ch http://www.lit-verlag.ch

Distribution:

In the UK: Global Book Marketing, e-mail: mo@centralbooks.com
In North America: Independent Publishers Group, e-mail: orders@ipgbook.com
In Germany: LIT Verlag Fresnostr. 2, D-48159 Münster
Tel. +49 (0) 2 51-620 32 22, Fax +49 (0) 2 51-922 60 99, e-mail: vertrieb@lit-verlag.de

e-books are available at www.litwebshop.de

Table of Contents

Introduction

"Are national societies sufficiently mature to respect the day-to-day implementation of the territorial concepts of each of the peoples that form an integral part of the national whole? Is interculturality applicable to the sphere of territoriality?" (Surrallés, García Hierro: 2005:9)

"The people here think that our culture is our dances, but culture is our way of being, the way we handle ourselves, our language, our walk"[i]

The Gran Chaco, the second largest biome of South America, entered a phase of deep and fast environmental changes a few decades ago. In the Paraguayan Chaco, which covers sixty percent of the country, these transformations resulted from the country's development model, which promotes rapid economic growth through soybean cultivation and livestock farming. Indigenous peoples are considered among those most affected by these unsustainable forms of production and environmental destruction due to their particular relationship with and reliance on their lands, territories and natural resources.[1] In addition to Indigenous Organisations' and some NGOs' longstanding efforts to secure more land for present and future generations, indigenous peoples' exercise of territorial use rights and management of lands secured in the past decades are now challenged by deforestation and access loss. This dissertation focuses on how one indigenous group of the Paraguayan Chaco, the Angaité of La Patria, is responding to this challenge.

The magnitude of the environmental changes considered in this dissertation is reflected in official reports, academic research and media articles. The Gran Chaco, in general, and the Paraguayan Chaco, in particular, experienced incredibly high deforestation rates over the past two decades. Hansen et al. revealed that the deforestation dynamics in the Chaco

[i] *"Ko'ápe la gente oimo'ã cultura ha'e jeroky, pero cultura ore reko, ore manejo, ore ñe'ẽ, ore guata." (Mr. Damasio Flores, November 2015)*

woodlands of Paraguay, Argentina and Bolivia from the year 2000 to 2012 represented one of the highest rates of tropical forest loss worldwide.[2] The soy boom experienced in the Eastern region of Paraguay in the last decades indirectly impacted the Chaco area because the cattle ranches from the first relocated to the latter. The expansion of old cattle ranches and the development of new ones are the main cause of the high deforestation rates experienced in the Paraguayan Chaco in the past years.

Therefore, the current situation of the indigenous peoples in the Paraguayan Chaco is a matter of concern. Though the Paraguayan constitutional framework recognizes indigenous peoples' rights it does not ensure they can be exercised. According to the UN Special Rapporteur on the rights of indigenous peoples, in Paraguay, there is a widespread lack of legal protection of rights to lands, territories and resources *"which are vital to ensure their survival and uphold their dignity."*[3] In particular, the Rapporteur mentions the existence of many cases regarding indigenous peoples' lack of access to and control over what they are legally entitled to. In addition, the UN Special Rapporteur on contemporary forms of slavery, recently raised her concerns about the working conditions of indigenous peoples in cattle ranches. There they are *"subjected to practices that would constitute forced labour and a form of contemporary slavery."*[4]

The new challenges faced by indigenous peoples are added to the longstanding struggle of restoring their historically dispossessed territories. In this situation, scholars as well as indigenist[ii] individuals and institutions tend to raise the following question: What should be done now? In response, some anthropologists argue that the indigenous lands legitimised in the past decades were unacceptable both in terms of quality and quantity. They add that the legal recognition of territories was clearly not enough to challenge the colonial mentality with which national societies relate to indigenous territories.[5] Indeed, indigenous peoples' rights to land are formally recognized in the Paraguayan Constitution. The following statement evidences the latter claim, "*The indigenous peoples have the right to communal ownership of the land, in an extension and quality sufficient for the preservation and the development of their particular ways of life.*"[6] However, recent research on the

[ii] *The term "indigenist" should not be confused with "indigenous". Following Diaz-Polanco (1997:23) and Blaser (2004:68) I use "indigenist" to refer to non-indigenous institutions and individuals that promote practices and policies that target indigenous peoples.*

implementation of this indigenous right indicates that the minimum of 100 hectares per family is not achieved in the Chaco region.[7]

It has been suggested that Paraguayan indigenist individuals' and institutions' approach to changing the indigenous peoples situation divides them into two distinct sectors.[8] According to Mario Blaser's argument, a particular conception of indigenousness and the needs associated to it is what both separates and connects these two sectors. The core distinction between these groups is whether they conceive of indigenousness as associated to a hunting-gathering way of life and the authentic traits these are supposed to display. The first sector, the "conservative", argues that indigenous people are not hunter-gatherers anymore and should be *"incorporated into civilization"*. By the latter they mean that development should be pursued without touching the current structure of land tenure. The other sector, the "radical", proposes profound reforms of the land structure in order to secure enough land to allow them to continue their hunter-gatherer way of life.[9] This scholar's main argument is that the two sectors fighting to impose what indigenous peoples are, left the indigenous voices out of the debate and lost sight on the fact that communities are complex emergent processes. The classification of Paraguayan indigenists into two camps might be criticised for oversimplifying a complex field. The sector identifies as "radical" encompasses considerable distinct positions and practices that would justify a separate treatment and analysis. I must admit that due to my personal history and professional involvement with institutions and projects that would be considered "radical", Blaser's main argument destabilised my basic assumptions about indigeneity and the indigenous territorial struggle. My first reaction was to try to justify a strategic use of the hunter-gatherer argument considering the terrible current situation of the indigenous people in Paraguay. Borrowing Valentina Bonifacio's[10] words on a related critique, I asked myself: If we leave aside the "essentialist" and "radical" indigenist and the "paternalistic" NGOs, who might be the allies of indigenous people in a region increasingly controlled by powerful landowners and drug traffickers? This dissertation is an exploration fuelled by the vacuum which the critique of Paraguayan indigenist individuals and institutions and the broader ranging indigeneity critique created.

While collaborating with two indigenist Paraguayan NGOs I noticed their growing concern for the behaviour of indigenous leaders and/or indigenous organisations that, from a non-indigenous perspective, seemed contradictory. Anthropologists with years of experience in collaborative projects with indigenous peoples and organizations share this apprehension. They highlight the occurrence of apparently inconsequential behaviour as well as contradictions in how indigenous people and external collaborators interpret specific problems or circumstances.[11]

The implications of the indigeneity critique, as well as the recognition of contradictions and misunderstandings in collaborations with indigenous peoples and organisations, determined the focus of my research. Instead of answering the above-mentioned question, "what should be done?", I explored how the Angaité themselves understand, explain and respond to the changes they face. To deepen my analysis I concentrated mainly, but not exclusively, on changes related to natural resources inside and outside one specific indigenous community, La Patria.

This dissertation aims to explore the following questions: How did the Angaité of La Patria respond to changes in access to, use and management of natural resources inside and outside their colony during the past 20 years (1995-2015)? For analytical purposes, this main question has been divided into three sub-questions: How did their livelihood strategies and land use patterns change during this period? How is this process of change in human-environmental relations reflected in the contradictions, tensions and internal political conflicts that mark everyday life in La Patria? How did external actors empower the Angaité to strengthen their livelihood strategies and manage their territory in accordance with their own priorities in order to preserve and develop their particular way of life?

I was aware, when I formulated these questions, of the implications and limitations of the "natural resources" concept. Indigenous understandings of and relations to territories transcend this "naturalist" concept, based on the Western Nature-Culture ontological divide. Considering this, some political ecologists conceive some resource use and territorial struggles as ontological conflicts.[12]

Nature's influence on culture has always been explored in Anthropology, even long before the Environmental Anthropology subdiscipline was established. Human-environmental relations became a central issue with Julian Steward's Cultural Ecology. Since then, different ways of conceptualizing the Nature-Culture interface have been developed within anthropology. Critically rethinking the main assumptions implicit in the notions of culture, society and nature on which previous anthropological approaches were based. A new set of perspectives on the study of human-environmental relations emerged from the Political Ecology umbrella. The first-generation of political ecologists achieved a more complex and balanced understanding of human-environmental relations but did not question the Western ontological dualism. The second-generation of political ecologists accepted the social construction of nature. They described how different cultures arranged contrasting meanings or representations of nature to show how conflicts related to these different representations.[13] Political Ecology initially reproduced the Western Nature-Culture division. A third-generation of political ecologists seeks to revert this by proposing an approach that considers both the implications of the ontologically specific character of biophysical reality and the existence of multiple ontologies.[14] The political ontology project, developed among others by Mario Blaser, Arturo Escobar and Marisol De la Cadena, constitutes one of these attempts.

The Angaité, like many other indigenous groups, have a history of creative engagement with the natural, social, local and global environment. Angaité families that were resettled in La Patria colony around 1983 creatively adapted their livelihood strategies. This constitutes another stage within their long process of territorial dispossession and concomitant integration into the Paraguayan society and market economy. Their "new" livelihood strategies as well as the territory and resources utilized were described by the British anthropologist Andrew Leake in 1995. That year, the land extension used by the Angaité of La Patria was ten times bigger than the land bought for them by the Anglican church. The Angaité had and have the constitutional right to use areas and resources traditionally accessed by them but owned by third parties. I refer to this territory as "La Patria territory".

But, what size and quality of land would suffice for the Angaité of La Patria to preserve and develop their particular way of life? Leake argued that land should be assessed in relation to how it strengthens and develops indigenous households' current livelihood strategies in accordance to their priorities.[15] He underscores that contemporary indigenous livelihood strategies tend to draw on resources and opportunities over broad regions. Therefore, any land-use restrictions should be considered as a violation of their sovereignty and self-determination.[16] In the same line, other scholars argued that one of the main factors that challenge the full enjoyment of indigenous peoples' *modus vivendi* is the current superimposition of other spaces on indigenous territorial space.[17]

In his book *Amazonia*, Paul Little introduced the cosmography concept to refer to *"the collective, historically contingent identities, ideologies and environmental knowledge systems developed by a social group to establish and maintain a human territory."*[18] Little defines territory as the "*appropriation of a specific portion of a region by a social group according to their respective cosmographical tenets.*"[19] This concept is associated to second generation political ecologists. However, I argue that cosmography remains useful when updated and inserted in current political ontology and anthropological debates. My conception of the cosmography is based on Arturo Escobar's contribution on "nature regimes", Ingold's "dwelling perspective" and on Tytelman's approach to the "place practices."[20] I use the concept to refer to the changing ideas and practices that contribute to the production of a particular place or territory; therefore, to the production of a particular understanding of the world. In this dissertation, I analyse the Angaité responses as changes in cosmographical practices. I focus mainly on two types of practices: ritual practices, like the *choqueo* and shamanism, and livelihood related practices. Following Tytelman, I conceive "La Patria territory", "La Patria colony" and the cattle ranches as three *"distinctive patterned places, which are produced by the practices of heterogeneous assemblages of humans and non-humans."*[21] Throughout this dissertation, I interrogate my own empirical data about the Angaité of La Patria by placing them in dialogue with diverse anthropological theoretical approaches. I argue that Blaser's political ontology, combined with the contributions of other anthropologist working on the Gran Chaco and lowland South America provides a useful theoretical approach to understand the responses of the Angaité.

Dissertation organisation

In **Chapter 1** I delineate a theoretical framework to understand the Angaité responses. I engage with the political ontology project to try to overcome the tendency to separate the human-environmental relations into a dichotomy. The first is the cosmological construction or representation of the environment. The second is the people's practical interaction with environmental resources in the context of subsistence activities.[22]

In **Chapter 2** I make a general contextualization of the Angaité and La Patria colony. La Patria colony has 22.520 hectares and is one of the seven Angaité communities that currently exist in the Paraguayan Chaco. Some Angaité families were resettled there around 1983, when the land was purchased by the Anglican church for that purpose. These families previously lived in small groups in areas later sold to private companies by the Paraguayan state at the end of the 19th century.

The Angaité have once again, since 1995, been facing a process of territorial dispossession. Despite the national and international legislation that granted the Angaité the right to access, use, and manage the natural resources of their territory. In **Chapter 3,** I explore how, by *de facto* not recognising "La Patria territory", the Paraguayan state allowed and promoted resource use practices that were incompatible with certain Angaité livelihood practices.

In order to understand how the Angaité of La Patria responded to these practices, in **Chapter 4**, I examine and describe how their livelihood strategies changed in the past two decades. I argue that the Angaité responded to these changes by adapting their livelihood strategies and land use. They abandoned some practices and adopted or intensified others. I claim that the Paraguayan state promoted incompatible resource use practices in the ranches which forced the Angaité to change certain livelihood practices and affected the production of "La Patria territory." Cosmographical practices contribute to the production of a particular place/territory.

In **Chapter 5**, I investigate how external actors succeeded or failed to empower the Angaité. The aim was to empower the latter to strengthen their livelihood strategies and manage their territory in accordance with their own priorities in order to preserve and develop their particular way of life. The inspiration for this chapter was the implementation and formulation of the indigenous peoples Paraguayan constitutional right, which I cite above. I argue that the simultaneous adaptation of livelihood strategies and the new patronage networks with the state/NGOs impacted the Angaité cosmography, leadership and networks of sharing. Thus, it redefined the "political" and the "social collective". In doing so, it also undermined the Angaité's capacity to manage their livelihood strategies and their territory in accordance to their own priorities.

Then, in **Chapter 6,** I analyse and describe the contradictions, tensions and internal political conflicts that characterize everyday life in La Patria. The redefinition of the "political" and the "social collective" clearly manifest in everyday contradictions, tensions and internal political conflicts. These tensions also reflect the preponderance of the Cartesian moral logic in detriment of the relational moral logic that traditionally characterised the Angaité cosmography.

In the conclusion, I return to my main arguments in order to answer my research questions. I argue that the responses from the Angaité should be understood as part of a longer and broader process of *adaptive resistance*[23] to sustain their own existence in their interaction with other worlds. The attempt to defend their practices and understanding of the world ends up reinforcing the Cartesian moral logic underlying the "development cosmography" of the Paraguayan state. The Angaité, conscious and worried about these changes and their disempowering implications, are trying to regain protagonism.

---

[1] Victoria Tauli-Corpuz, 2015
[2] M. C. Hansen et al., 2013
[3] Victoria Tauli-Corpuz, 2015
[4] Urmila Bhoola, 2017
[5] Alexandre Surrallés and Pedro García Hierro, 2005
[6] PNC, 2012
[7] Oscar Ayala, 2016
[8] Mario Blaser, 2004b
[9] Ibid.
[10] Valentina Bonifacio, 2017
[11] Volker Von Bremen, 2017
[12] Mario Blaser and Arturo Escobar, 2016
[13] Ibid.
[14] Ibid.
[15] Andrew Leake, 2006
[16] Andrew Leake, 2006
[17] Alexandre Surrallés and Pedro García Hierro, 2005
[18] Paul E. Little, 2001: 5
[19] Ibid., 246
[20] Arturo Escobar, 2016
[21] Carolina Tytelman, 2016: 18
[22] Tim Ingold, 2000
[23] Bruce Albert, 2005

1 Theoretical framework and methodology

1.1 Introduction

Conflicts over territorial dispossession involving indigenous people are often related to land, fishing and hunting rights, as well as opposing territorial concepts, systems of land use and ideas about how to manage natural resources.[1] In order to contextualize my research in broader theoretical debates, in section 1.2 I explore the emergence of Political Ecology as a critical approach to investigating human-environmental relations. In section 1.3 I explore how human-environmental relations, "indigenous people" and "territory," have been conceptualized. I first present two approaches to the study of human-environmental relations change in the Gran Chaco context. Secondly, I outline how the conceptualization of these terms evolved in the political field of the international and Paraguayan legislation. I conclude by briefly presenting the main elements of the indigeneity critique in order to review how Political Ecology is dealing with its implications. In section 1.4, I introduce a series of conceptual contributions from Political Ecology to the understanding of the Angaité responses and to the study of indigenous territorial struggles as ontological conflicts. Finally, in section 1.5, I posit methodological considerations that arose from this debate.

1.2 Political Ecology as an approach to human-environmental relations

As part of the Western scientific tradition, anthropological approaches to human-environmental relations tended, for many years, to lean towards either side of a dichotomy: the mythic-religious or cosmological construction of the environment and the people's practical-technical interaction with environmental resources in the context of subsistence activities.[2] In other words, until quite recently, the main anthropological approaches tended to be either mentalist/symbolist or materialist.[3]

The importance of nature and its influence on culture was central to the discipline long before its sub-discipline, Environmental Anthropology, was established. However, it was Julian Steward's Cultural Ecology that made human-environmental relations a core issue. This approach focused on the relationship between the subsistence economy of culturally defined populations and their natural environments. Cultural Ecology recognized the role played by the environment in shaping cultures, providing possibilities and setting constraints.[4] Denying the previous simplistic geographical determinism allowed Steward to define culture as an adaptive mechanism. In short, the cultural ecological perspective viewed culture as expressing collective human engagement with challenges and opportunities posed by local environments.[5]

Since then, Anthropology has developed different approaches to conceptualizing the nature/culture interface. These approaches resulted in social and political implications, but also, in most cases reproduced uncritically the Western division of nature and culture as separate realms.[6] It was rather recently that Political Ecology highlighted "*the interwoven character of the discursive, material, social and cultural dimensions of human-environmental relations.*"[7] The main assumptions implicated in the notions of culture, society and nature, on which previous anthropological approaches were based, were critically rethought. Thus, a new set of approaches to the study of human-environmental relations emerged under the Political Ecology umbrella. Political Ecology initially reproduced the Western nature/culture division. However, recent contributions are reshaping the approach in order to consider the existence of multiple ontologies.

This section briefly explores the emergence and evolution of Political Ecology, a central approach in my theoretical framework. This field of study emerged in the 1970s to address the interrelation between nature, culture and power. It integrated an ecological and political economy frameworks to remedy Cultural Ecology's lack of attention to power and political economy's weak conceptualization of nature.[8] Political Ecology combined three major theoretical stands: Cultural Ecology's attempts to link the physical environment with cultural identity, ecological anthropology's cultural politics of adaptations and Marxist-inspired concern for the economic conditions of peasants.[9] Since then, this approach has been challenged and altered by social theory. The latter constitutes the series of discourses that seek to reflect on the constitution of social life to make social practices intelligible.[10] These epistemological and ontological discussions resulted in three differentiated but overlapping generations of political ecologists.

1.2.1 First generation of political ecologists

To tackle the weakness of previous frameworks, this first-generation of political ecologists resituated the question of adaptation "*into the larger scale in which world-system and dependency theory operated.*"[11] In his 1972 classic article, *Ownership and Political Ecology,* Eric Wolf posited that local rules of ownership and inheritance should be understood as "*mechanisms which mediate between the pressures emanating from the larger society and the exigencies of the local ecosystem.*"[12] From this perspective, local cultures not only adapt to the "*natural environment*" but also "*rework themselves creatively*" to cope with the pressures of the capitalist system and meet their own ends.[13] Scholars used the integration of Cultural Ecology and political economy to describe indigenous societies and cultures in their regional as well as historical context.[14]

Political ecologists considered three assumptions of equal importance: costs and benefits associated with environmental change are distributed unequally among actors; this unequal distribution of cost and benefits reinforces or reduces existing social and economic inequalities; environmental changes have differentiated social and economic impacts, both alter the power of actors in relation to other actors and the ability of the first to control or resist the latter.[15] In short, "*any change in environmental conditions must affect the political and economic status quo, and vice versa*" and the ubiquity of conflicts over environmental resources "*highlight the importance that diverse actors attach to those resources.*"[16]

This first-generation of political ecologists achieved a more complex and balanced understanding of human-environmental relations. However, they reproduced the Western ontological dualism by maintaining "nature" or the "environment" as distinct domains from "culture".[17] A second critique was that this generation of scholars retained a strong commitment to epistemological realism, "*assuming that reality was accessible from the standpoint of the sciences.*"[18] A second-generation of political ecologists entered the theoretical debate by scrutinizing the first's assumptions.

1.2.2 Second generation of political ecologists

Influenced by poststructuralist, post-colonialist and feminist theoretical trends, as well as epistemological debates fostered by constructivism and anti-essentialism, a second-generation of political ecologists emerged in the last two decades.[19] This second-generation critiqued the epistemological assumptions of their predecessors. They entered the essentialist/constructivist debate by addressing the relation between knowledge and reality, thought and the real.[i] From a diverse series of epistemologies of nature these political ecologists accepted the social construction of nature without denying that there is a real nature. They insisted on the necessity to discursively analyse modern forms of knowledge.[20] Political ecologists posited that the analyses of the following systems were inextricably bound: systems of meanings of nature, systems of use of resources, systems of production and systems of signification.[21]

With its emphasis on discourse, representation and imagery, constructivism provided the conceptual root to evaluate the perceptions and rationality of land users, government officials, conservationists and scientists.[22] Constructivist political ecologists, mainly anthropologists or geographers, described how different cultures arranged contrasting meanings or representations of nature. They showed how conflicts related to these differences as well as how meaning and representation relates to what is actually done to nature.[23] Geographers generally focused on deconstructing notions of "nature" and anthropologists studied "ethnic identities" or notions of "culture".[24] Paul Little's Political Ecology, with the cosmography concept at its core, is representative of this generation.

The critique against constructivist political ecologists is that they *"tend[ed] to bypass the question of the ontologically specific character of biophysical reality (…) treating material entities as mere envelopes of meaning (…) [which] contributes to the continued dominance of epistemological realism."*[25] Tim Ingold, who is interested in the biophysical contribution to human societies and whose "*dwelling perspective*"[26] is useful to understand livelihood change in societies that were/are traditionally hunters and gatherers, heeded this critique.

[i] *"Succinctly, essentialism is the belief that things possess an unchanging core, independent of context and interaction with other things, which knowledge can progressively know. The world, in other words, is always predetermined from the real. Constructivism, on the contrary, accepts the ineluctable connectedness between subject and object of knowledge and, consequently, the problematic relation between thought and the real. From a certain constructivist perspective (Foucauldian and Deleuzian poststructuralism in particular) language and discourse do not reflect "reality" but constitute it." Mario Blaser and Arturo Escobar, 2016: 167*

1.2.3 Third generation of political ecologists

This third-generation of political ecologists is not only inspired by academic debates that claim new shifts in social theory. They are fuelled by the visibility of some environmental conflicts that cannot easily be analysed from a realistic account of struggles over "nature" or "resources" and the evidence of persistence of worlds not reducible to a universal modernity.[27] What differentiates the third from the second generation of political ecologists is their shift in focus from epistemology to ontologies. In summary, instead of rethinking the relation between "nature" and "culture", this emergent line of thought is challenging the ontological character of the divide, as mentioned, usually taken for granted in Environmental Anthropology as universally applicable.[28]

The main intellectual sources of this debate are: phenomenology, Actor Network Theory, Deleuzian philosophy and ethnographies of non-Western groups.[29] Anthropologists contribute novel approaches to nature/culture questions. Their predecessors are Tim Ingold, Marilyn Strathern and Philippe Descola, among others.[30] Rubenstein's *Poststructuralist political ecology of the Amazonia* could be situated in the transition from the second to the third generation. This scholar argued that the task of Political Ecology was to "*analyse spatially distributed fields of power, without privileging the perspective of one agent in this field.*"[31] One of his aims was to expose how persistent binaries operate (nature/culture, savage/civilized, traditional/modern) and often influence our understanding of indigenous people and their environments.[32]

Current academic debates on post-representational epistemologies and relational ontologies are returning to questions about "the real" related to the current neo-realist or post-constructivist trend in social theory.[33] It is still unclear how neo-realism might contribute to a different understanding of the cultural dimension of nature/culture regimes. Yet it remains one of the driving motivations of political ontology. It is one of the various approaches labelled as pertaining to the third-generation of political ecologists.[34] I will come back to political ontology, but first let me introduce two perceptions of cultural change and human-environmental relations in chaquean anthropology.

1.3 Human-environmental relations, territoriality and the indigeneity critique

1.3.1 Contrasting approaches to cultural change and human-environmental relations in chaquean anthropology

Chaquean anthropologists have been debating how the colonization of the Chaco and the dispossession of territories impacted the human-environmental relations of the indigenous people of the Chaco, for decades. To illustrate two approaches to this discussion, that had and have influenced the chaquean anthropology, I selected two scholars: Volker Von Bremen and Gastón Gordillo. The main difference between them is their conception of culture and the role attributed to it in indigenous people's responses to the problems faced in their new situation. The local models of human-environmental relations and the role this model plays in the interaction between indigenous people and external actors, the latter include the state, NGOs, missionaries and anthropologist, is also central to this debate.

Von Bremen argued, in the 80's, that indigenous people's behaviour and responses to the new challenges evidenced the persistence of a hunter-gatherer mentality. He posited that the responses of the indigenous people of the Chaco to new challenges, which they faced as a consequence of the destruction of their traditional way of life, were based on the logic that underlines this groups relation to "nature".[35] Von Bremen described some characteristics shared by the indigenous peoples of the Chaco in order to clarify his argument. An important aspect he mentions is that for these groups, "*the conceptual division [of culture and nature] is not existent… For them, natural and cultural phenomenon have equal quality and origin.*"[36] Another shared characteristic, he states, is the belief that the natural resources for their survival are abundant. How to appropriate the already existent resources is important for social reproduction. Due to the central and persistent role Von Bremen attributes to the hunter-gatherer culture (as mentality, values), his approach has been criticised as culturalist or essentialist.[37] In order to present and defend political ontology as a new approach, Mario Blaser also criticises Von Bremen, including him in what he calls "*the hunter-gatherer paradigm.*"[38]

In his article published in 1993, "*La actual dinámica económica de los cazadores-recolectores del Gran Chaco y los deseos imaginarios del esencialismo*", the Argentine anthropologist Gastón Gordillo identifies himself as part of a new theoretical current that emphasises the historical processes through which hunter-gatherer societies have been incorporated into broader socioeconomic systems.[39] He includes Von Bremen among the scholars that oppose this current and "*argue that the hunter-gatherer culture of these groups translates in economic strategies that are determining their modalities of insertion to the market.*"[40] Gordillo disavows that the failure of the development projects in the indigenous colonies and the economic strategies in the new context are related to a hunter gatherer culture. Instead, he claims that they are related to the economic rationality behind these projects, "*that was in conflict with the domestic and reciprocal relations of the aboriginal groups whose minimum efficiency criterium is not to obtain an average profit but to achieve simple reproduction.*"[41] In short, for Gordillo, indigenous economic strategies are explained by the cumulative dynamics historically imposed by capitalism in the Gran Chaco.

The question remains: what role does culture, as a recurrently reformulated social and symbolic system, play in the current economic strategies of the indigenous people of the Chaco? In Gordillo's argument, culture's role is that of the weakened reciprocal and egalitarian ideology that affects their productive relations. Thus, it limits forms of wealth accumulation culturally considered excessive.[42] He succinctly defines their culture as "*a dynamic system in constant reformulation, that shapes actions and responses... but is simultaneously strongly conditioned by the limits imposed by capitalism.*"[43] In this line, indigenous people's specificity is a historical construct resulting from their cultural heritage and their identity. It is the social role assigned to them by capitalism. They constitute the domestic productive sector that can generate great surplus.[44]

Throughout this dissertation, I engage with this debate based on my fieldwork among the Angaité. Though these scholars may not consider themselves political ecologists, their contrasting positions share interesting similitudes with the debates among political ecologists. I argue that Von Bremen was misunderstood, probably because he employed concepts considered outdated today. However, I will also argue that anthropologists like Tim Ingold and Mario Blaser have, paradoxically, developed very similar arguments. In doing so, I incorporate new elements to this old debate. I include scholars that contribute different approaches and assist in better understanding the Angaité's responses to their current situation.

1.3.2 Territory and indigenous peoples' rights

The three main pillars of the international legislation concerning indigenous peoples' rights are: the International Labour Organization's (ILO) "Convention 169", the United Nations' "Declaration on the Rights of Indigenous Peoples" and the jurisprudence of the Inter American Human Right System.[45] This legislation is a product, among other influences, of the persistent lobbying of and for indigenous peoples' groups and organizations.

The ILO "Convention 107" is the predecessor of the "Convention 169". "Convention 107" on "Indigenous and Tribal Populations", passed in 1957, was the first to define "indigenous identity." This definition grouped heterodox people and places in order to bring them under the reach of international law.[46] "Convention 107" is part of an effort to situate the relationships between people and resources in relation to culture, the latter is the basis for a rights-based approach to development.[47] With regard to territories, it identified the need to "*guarantee access to indigenous people ´habitual territories´ in order to protect subsistence practices.*"[48]

The ILO "Convention 169" was formulated in the early 1970s by social scientists studying indigenous populations and became, by the 1980s, a key feature of international law.[49] Article 1.1 of this Convention defines indigenous peoples as *"peoples in independent countries who are regarded as indigenous on account of their descent from the populations which inhabited the country, or a geographical region to which the country belongs, at the time of conquest or colonization or the establishment of present state boundaries and who, irrespective of their legal status, retain some or all of their own social, economic, cultural and political institutions.*"[50] Therefore, indigeneity was defined as a "territory" composed of "*the total environment which the peoples concerned occupy or use.*"[51] Article 13, specifies that "*governments shall respect the special importance for the cultures and spiritual values of the peoples concerned of their relationship with the lands or territories.*"[52] This Convention already attributed an active role to indigenous people in the definition of their governance structures. However, some scholars argue that it is only within the "United Nations Declaration" that the right to self-determination was finally recognised.[53]

After two decades of negotiation, the United Nations' "Declaration on the Rights of Indigenous Peoples" was approved in 2006. The following is stated with regard to territories, "*Indigenous peoples have the right to maintain and strengthen their distinctive spiritual relationship with their traditionally owned or otherwise occupied and used lands, territories, waters and coastal seas and other resources and to uphold their responsibilities to future generations in this regard.*"[54] It also grants indigenous peoples the right to self-determination, and explicates that "*by virtue of that right they freely determine their political status and freely pursue their economic, social and cultural development.*"[55]

In parallel, and partly based on the above-mentioned documents, the Inter American Human Right System (IAHRS) developed a complex jurisprudence on the subject. An important report published by the IAHRS argues that a precise definition of "indigenous peoples" in international law is impossible due to the immense diversity of the indigenous peoples.[56] The IACHR refers to the "Convention 169" definition, adding that there are objective and subjective elements that define indigenous peoples. These include: historical continuity, territorial connection, as well as their own distinctive and specific social, economic, cultural and political institutions. The subjective element correspond to collective self-identification.[57] With regards to territories, the IACHR recognises that *"Indigenous and tribal peoples have unique ways of life, and their worldview is based on their close relationship with land. The lands they traditionally use and occupy are critical to their physical, cultural and spiritual vitality."*[58]

Paraguayan legislation on Indigenous Peoples' Human Rights is mainly contained in Chapter 5 of the National Constitution of 1992. It constitutes the following legislation: the Law 904/81 "*Estatuto de las comunidades indígenas*", the Law 234/93 that ratifies the ILO "Convention 169", the Law 253/93 that ratifies the Convention of Biological Diversity and the Law 1963/02 "*Que establece el Estatuto Agrario*".

In 1981, through the implementation of Law 904, the Paraguayan state policies towards indigenous people shifted. This change was provoked by the movement for indigenous rights related to the genocide of the Ache indigenous people and campaigns promoted, among others, by the German anthropologist Mark Münzel. The state defined the "indigenous communities" (*Comunidades indígenas)* as the socio-legal units that compose the different indigenous ethnic groups.[59] This law also established the procedure for recognition of communitarian leadership and legal personhood.[60] According to some scholars, the emergence of this law was the result of pressure from the NGO sector and two sets of strategic alliances. On one hand, between missionary and academic sectors. On the other, between important members of the army.[61]

The National Constitution recognises the existence of indigenous peoples, defined as "*groups of culture prior to the formation and the organization of the Paraguayan State.*"[62] Article 63 states: *"The right of the indigenous peoples to preserve and to develop their ethnic identity in the respective habitat is recognized and guaranteed. They have the right, likewise, to freely apply their systems of political, social, economic, cultural, and religious organization, as well as the voluntarily subjection to their customary norms for the regulation of [their] internal coexistence (convivencia)."*[63]

Regarding the right to territory, article 64 clearly states that *"The indigenous peoples have the right to communal ownership of the land, in an extension and quality sufficient for the preservation and the development of their particular ways of life. The State will provide them gratuitously with these lands, which will be non-sizeable, indivisible, non-transferrable, imprescriptible, not susceptible to guarantee contractual obligations nor to be leased; likewise, they will be exempt from taxes."*[64]

1.3.3 Indigenous peoples' rights and the indigeneity critique

In their critique of the grand narratives of anthropology, post-modernist and post-colonialist scholars argue that anthropologists have inadvertently served colonialism's aim, the hegemony of modernist ways of thinking and programmes of development.[65] More specifically, they claim that in its representations of other cultures anthropology has homogenised, exoticized and silenced real beings that lead particular lives.[66] Anthropology's cultural relativism has not remedied the use of Western classifications (science, religion, economics, politics, kinships, society, etc.) that carry a hidden homogenizing agenda.[67]

In the introduction, I paraphrase Blaser's suggestion that the Paraguayan indigenist field shares a particular conception of indigenousness associated with a hunting-gathering way of life. In this line, Emily Yeh and Joe Bryan analyse the uses of indigeneity as a term, concept and practice.[68] They pay special attention to the political and intellectual work it performs. From this perspective, anthropology contributed to the consolidation of "*colonial stereotypes of indigenous peoples as living at one with nature.*"[69] According to these scholars the characterization of indigeneity was based on a particular relationship to nature and place. This portrayal became part of Cultural Ecology's legacy which lends itself both to Political Ecology and international law. By turning to traditional societies to support their claims, Cultural Ecology "*bore a lasting impact on the understanding of these cultures as indigenous, their traits and practices expressly tied to particular environments and places.*"[70]

They argue that anthropologists, in this case cultural ecologists, played an influential role in drafting the ILO Convention 107 and Convention 169, providing an underlying logic to group societies that were radically local. The legal definition of territory used, aimed to "*secure a place for indigenous people within the configuration of modern political space.*"[71] These scholars trace the continued influence of Cultural Ecology's understanding of the relationship between indigenous peoples and their environment as systematic, the social expression of which constitute a "territory".[72] In other words, the survival of indigenous cultures depended on maintaining a relationship with the environments in which they had evolved. Yeh and Bryan relate Convention 169's distinction of peasant from indigenous/tribal population to agrarian studies' contrast between peasants, placed within capitalism, and indigenous peoples, living outside or at the frontiers of that system.[73]

The question remains: How wrong was Julian Steward's understanding of the relation between indigenous cultures and their territories? Apparently, the maps used by cultural ecologists were close to peoples' understanding of their own societies. In fact, many indigenous groups currently rely on these maps today as a record of their traditions.[74] In a similar process, from the 1990s, different groups around the world were adopting the term "indigenous people" to articulate their own claims to land, resources and self-determination.[75] Groups around the world developed awareness of "indigenous" identity due to the political necessity of employing this term.[76] In short, universal notions of indigeneity, highly criticised by social scientists, were inscribed in international law and continue to condition the recognition of indigenous peoples' rights.[77]

Discourses that represent indigenous people as groups with a strong connection to nature were and are used by indigenous groups. Rubenstein claims that the Kayapó exploit such discourse to create political capital though non-indigenous groups can use this discourse against them. He goes on to argue that the Kayapó's desire for manufactured goods is a sign of corruption or ethnocide.[78]

Other anthropologists, geographers and political ecologists were influenced by poststructuralist theories and modes of inquiry, like Yeh and Bryan. They also deconstructed and destabilized the concept of indigeneity in order to understand how discursive constructs, based on the ontological division of nature and culture, privilege and marginalize subjects.[79] In this critique, indigeneity is understood as an essentialist claim grounded in the belief that certain qualities are inherently indigenous.[80] The claim is that the conflation of indigenous people with the past and with nature is a social construct. It perpetuates European differentiations between savagery and civility as well as rationalizing the colonization of indigenous lands.[81] In this sense, the postcolonial critique helped identify the material and intellectual tools of dispossession of indigenous peoples from their lands and culture. The production of spaces, like the Indian reserves, is linked to the colonial apparatuses that legitimated the land dispossession.[82]

These critiques revealed many of the direct political implications of essentialist notions of indigeneity. Scholars claim that exoticism gives impetus to development projects: implicitly "inferior" forms of life assimilate "superior" people's knowledge to live a "better life".[83] In some cases, such as Amazonia, indigenous people are only recognized by the government agent if they display *authentic* cultural practices.[84]

The parallel emergence of indigenous people's movements and concerns regarding environmentally friendly forms of development led to the assumption that contemporary indigenous environmental practices always align with Western environmentalism, producing what is known as the "Ecologically Noble Savage".[85] It was assumed that protecting indigenous cultures would result in the protection of nature.[86] Current environmental practices reflect the historical dispossession and change processes that indigenous people, like everybody else, experience. In many cases their practices are characterised as more sustainable.[ii] However, the naturalist roots of Western environmentalism in many ways contradict the ways indigenous people conceive and relate to their environments. Again, indigenous people are accused of being contaminated by a "loss of culture" which mismatches representation and actual indigenous practices.[87]

Considering the academic and political use of the term "indigeneity", some scholars warned that this deconstruction can increase marginalization of indigenous people by denying that they have a distinct way of life.[88] They suggest that a strategic essentialism is justified and necessary to defend current indigenous claims. The judicious use of the idea of indigenous harmony with nature is employed to generate symbolic capital in the environmental arena for the advancement of political goals.[89] Rejecting this, it has been claimed that further marginalization could result from this well-meaning strategy.[90]

[ii] *see Berkes 1999; Brosius et al. 1998; Escobar 1995; Posey 1985*

I argue that "indigenous people" should and will continued to be used, more so now that it has become part of a political identity. When I use the term I connote the Paraguayan law's definition, as "*groups of culture prior to the formation and the organization of the Paraguayan State.*"[91] I do not assume any essential difference that allows a clear delimitation of who is or is not indigenous. The following definition is included in the Paraguayan National Indigenous Population Census of 2002 and 2012: An indigenous individual belongs to an ethnic or indigenous group as well as a community, group of families or indigenous neighbourhood, regardless of his use or loss of his native language.[92] I use the term "indigenous communities," following the terminology of the 2002 Census, to refer to entire indigenous landholdings, and "villages" as its internal settlements.[93] Sometimes I use the term colony, also used by the Angaité themselves, as synonymous to community.

1.4 Understanding Angaité responses from a third-generation political ecologists' perspective

The first generation of political ecologists drew on agrarian studies to critique cultural ecologist's conceptualization of indigenous societies as living close to nature, or as living outside of modern forms of economy and power.[94] These early political ecologists portrayed the indigenous as peasants in addition to analysing the social inequalities and marginalization they faced, a result of their integration in capitalist markets.[95] The second generation of political ecologists injected cultural politics into environmental debates. They actively used indigeneity as a means of articulation of experiences of dispossession and assimilation into modern societies. These political ecologists turned indigeneity into an explicit political identity.[96] Some scholars among the third generation of political ecologists tried to overcome the division of the world in two types of cultures (traditional and modern). They follow Bruno Latour's proposition of a symmetrical science that treats nature and culture as well as indigenous and Western societies equally.[97]

Despite the fact that the new ways of thinking about identity, difference and nature sits uneasily with some existential and political claims of indigenous people to being, they attempt to dismantle restrictive expectation about authenticity and tradition in the constitution of indigeneity.[98] Arturo Escobar explicitly rejects the view of indigenous people as quintessentially organic and embedded in nature. He suggests launching a theoretical and political defence of the "organic nature regime" as a historical regime.[99] Escobar's "nature regimes" and Blaser's "political ontology," which I address in the following paragraphs, can be situated in the context of the above-mentioned new approaches

1.4.1 Political ontology

The critique of the indigeneity concept has led scholars to develop new ways of thinking about identity, difference and nature. These approaches emphasize movement, disconnection and hybridity in contrast to a rooted notion of essential identity.[100] The political ontology approach, developed among others by Mario Blaser, Arturo Escobar and Marisol De la Cadena, is situated in this context. Political ontology is one of the new frameworks that emerges from the convergence of indigenous studies, science and technology studies (STS). Post-humanism and Political Ecology helps us understand ontological conflicts which involve different suppositions of "what exists".[101] In order to consider ontological conflicts as sites of potential, Blaser argues, that the fundamental assumption of the social sciences must be questioned. The supposition is that we are all modern and differences lie between cultural perspectives of a single reality "out there".[102]

On one hand, political ontology refers "*to the power-laden practices involved in bringing into being a particular world or ontology.*" On the other hand, it "*focuses on the inter-relations among worlds, including the conflicts that ensue as different ontologies strive to sustain their own existence in their interaction with other worlds.*"[103] This approach directs the attention "*to the more-than-human forces that shape lifeworlds*" and to the "*different ontological commitments and ways of worlding.*"[104] By doing so, it suggest a widening of the concept of politics and of the claims that nature, as an universal analytical category, may be as indispensable as it is inadequate for liberatory projects.[105]

Blaser's political ontology framework has concrete implications in my research. His contributions to political ontology were outlined in his monograph[iii] and supplemented by further articles that proposed notions of "worlds", "ways of worlding" and "pluriverse".[106]

Blaser's definition of ontology has three layers. The first layer comprises the implicit or explicit assumptions of what kinds of things do or can exist. As well as the latter's conditions of existence and relations of dependency.

[iii] *see Blaser 2003 and Blaser 2010*

The second, borrowed from Actor Network Theory (ANT), considers that "*ontologies do not precede mundane practices, but rather are shaped through the practices and interactions of both humans and nonhumans*".[107] Ontologies perform themselves into worlds and are considered synonyms. The third layer traces the connection between myths and practices. Ontologies manifest as stories and myths. To grasp what kind of things and relations make up a given world the verbalized and enacted (embodied) aspect of myth should be considered.[108] Modernity is, for example, understood as the enactment or state of being of the modern myth or ontology.

Considering the Angaité's territorial struggle as an ontological conflict implies the recognition of their world as an existent alternative to modernity. The Angaité's responses to this territorial struggle are part of a pursuit for their own life projects. Blaser defines these life projects as *"based on visions of a good life premised on densely and uniquely woven threads of ontological assumptions, the materiality of landscapes, memories, practices, expectations, and desires."*[109]

Based on the contributions of different scholars, Blaser contextualises the emergence and main characteristics of the "modern ontology". He states, that between the 16th and 17th centuries, a new way of conceiving truth and knowledge began to emerge in certain parts of Europe. These new conceptions were based on a disjunctive view of language. This disjuncture was akin to the distinction between mind and world that characterises the Cartesian dualism. Thus, truth came to be conceived as "*the coincidence of word and world* (…) *the former is a completely adequate and transparent representation of the latter.*"[110] Cartesian dualism and the system of dualistic sets such as society/nature, mind/world, word/world, are elements of the modern myth or ontology.

The emergence of this ontology took place along with the new relations between the "Indians", non-moderns or the people of the New World.[111] This is important, because "Indians" (today called indigenous people) are part of the modern ontology since its constitution. Another important element of the modern ontology is the notion of linear time in form of the evolutionary dynamic. The three main mutually constitutive elements of the modern ontology which developed since then are: nature, Indians/non-moderns and development.[112] Nature, as opposed to society "*has been conceived as an out of (human) history, an autonomous domain operating as a source of truth, authenticity and morality.*"[113]

Figure 1 sketches how these three elements relate in the modern myth. There is a clear ontological distinction between nature and culture, a clear tendency to conceive difference in hierarchical terms and a linear conception of time. The domain of culture is subdivided into several cultures and the place of "modern culture" is linked to a hierarchical background based on linear time. Thus, modernity tames the different ontologies in order to shrink ways of worlding and make them fit into the squares labelled "other culture".[114]

Figure 1. The modern myth (Blaser 2013:554)

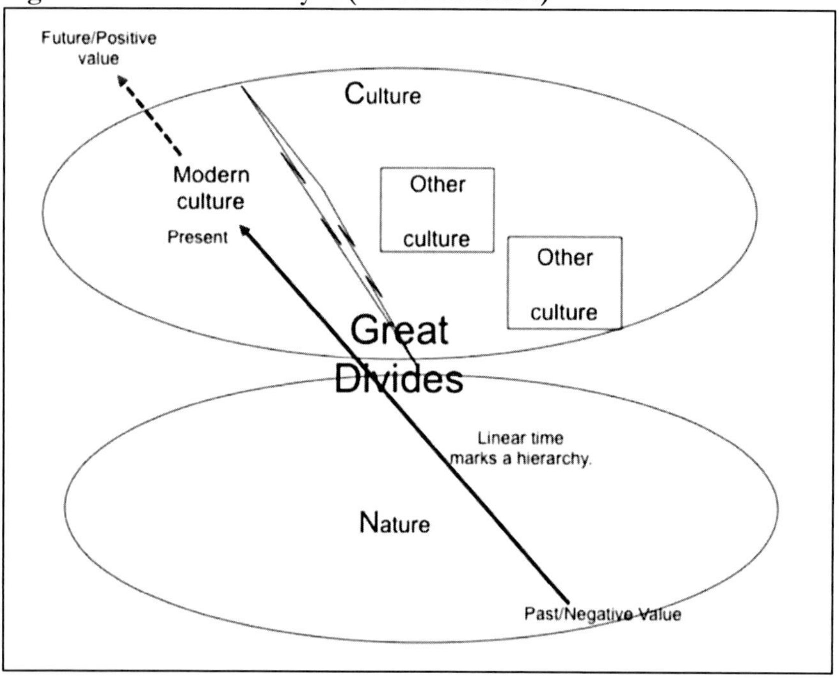

Radically different ways of worlding emerge from a process that connects other ways of worlding. Political ontology speaks of ontological differences in contrast to cultural differences. Blaser states that this responds to an analytical decision regarding how to address difference, not kinds of differences. In fact, what the political ontology project tries to do is show the ontological nature of the language of culture to refer to differences. This language is dominant not only in academia, but among the subjects of academic analyses, "*who render ways of worlding into culture and mobilize it in practices of belonging (identity).*"[115]

Blaser applied and further developed this approach based on his experience with the *Yshiro* people of the Paraguayan Chaco. In his Doctoral dissertation, he explores the "*epistemological and political possibilities of the relational moral logic implicit in some Yshiro intellectuals' discourses and conduct*".[116] The *yrmo,* or *Yshiro* world exemplifies a radically different world enacted before us. It involves computers and internet along with other nonhumans that are older and changing.[117] He argues that the *Yshiro* non-dualist ontology is characterised by a *relational moral logic.* It is opposed to the *Cartesian moral logic* that characterises the Western/modern ontology. While the relational moral logic gives precedence to relations over entities, he argues that the Cartesian moral logic gives precedence to self-referential entities over relations. The opposition of the *Yshiro* ontology and the Western ontology should not be understood as a homogenization of the multiple experiences. Ontologies or the ways of worlding of indigenous people exist independently of those of the modern West. Political ontology does not ignore variations or dissent within modernity. Instead, it shrinks modernity to analytically focus on other ways of worlding that exist. This approach is concerned with practices, not with specific groups. Then, the mere indigenous involvement does not create a conflict, an ontological one.[118]

As Blaser states, ontological conflicts pose the challenge of how to account for them without re-enacting the ontological assumption of a reality "out there" being described.[119] How did Blaser solve the problem of using modern terminology, historically charged with meanings derived from the modern ontology, and at the same time dialogue with the *Yshiro* ontology? Blaser's solution was to use the language developed in science and technology studies (STS) and Actor Network Theory (ANT). According to this idiom, reality does not precede practice but is rather shaped within these practices.[120] He uses the term *actants* to refer to entities that "*are constituted only in and through their mutual relations and not by any intrinsic property.*"[121] He uses this term to try to move beyond problematic dichotomies, such as living/non-living entities. This implies that in his narration, all elements were taken to be symmetrical, with no restrictions in terms of agency or causality. In fact, there was no hierarchy among entities and they often play identical roles.[122]

Using this framework Blaser reads and criticises Von Bremen's hunter-gatherer mentality argument as part of the "hunter-gatherer paradigm". This paradigm makes a basic distinction between indigenous and non-indigenous people based on a contrasting essential morality that exists outside history.

Blaser makes clear that he is not opposing the idea that egalitarism is a value that might be embedded in the practices of many indigenous people. Instead, he argues against *"taking this egalitarism as something inscribed in the being of indigenous societies and not as an effect of contingent and changing circumstances."*[123] He claims that this paradigm has been highly influential in the way indigenous people have been represented and consequently treated in Paraguay. He situates the scholars that developed this paradigm as part of expert development institutions that shape the regime of truth of modernity. Their role, he argues supported by Michel Foucault, is to define certain ways of imagining reality as knowledge while defining others as myth. Blaser claims, that the propositions from these scholars were moral judgments presented as facts. Following this argument, he points out that from their *situated moralities,* these experts were "*heavily imbued with the Cartesian moral logic, [and] tend to translate practices embodying relational moral logics into immutable traits.*"[124] In other words, the way indigenous people have been portrayed in this paradigm resulted from particular positions and complex/mutually contesting demands. As an example, Blaser mentions the contesting demand of related clear-cut categories usable in the context of development projects. The commitment to cultural diversity was resolved by a list of traits that define an authentic indigenousness as really different from "Westerness".[125]

Instead of focusing on mentalities, Blaser approaches development as a practice and discourse that aims to universalize modernity and its institutions.[126] He contends that development materializes as practices, in Paraguay, through pervasive patronage networks. Throughout the monography development is defined as a circulation network. In other words, the modern regime or truth production, or attempted production of truth, is approached as an object of governance. He also considers the values and institutions through which these are governed. Last of all, Blaser analyses how these objects of government (the *Yshiro* as indigenous people) responded to and in turn transformed those institutions and values meant to govern them.[127] This approach, based on Latour's symmetrical anthropology, allowed him to see the networks or relations woven into life-giving webs that constitute the world or reality. By overcoming the "great divide", Blaser, also overcomes the "traditional-modern" societies separation.

Blaser adopts an understanding of reality borrowed mainly from ANT to bypass the nature-culture divide. Thus, he avoids the assumption that the reality is "out there" and that "in here" we have cultural representations of it. This understanding of reality also implies that reality is always in the making *"through the dynamic relations of hybrid assemblages that only after the fact are purified by moderns as pertaining to either nature or culture."*[128] Central to this understanding is the conception that humans are involved in the enactment of realities but are conditioned by previous actions of agents that are heterogeneous assemblages which we (moderns) classify as humans and nonhumans.[129]

In the following chapters, I adopt the political ontology framework to seek an answer to my research questions. However, at the same time I critically analyse this framework by exploring the differences and similarities shared with previous approaches.

1.4.2 Nature regimes

In *Steps to an Antiessentialist Political Ecology,* Escobar explores what he calls the decline of the modern ideology of naturalism, "*the belief in the existence of pristine Nature outside of history and human context.*"[130] Without denying the existence of a biophysical reality, he argues that nature is always constructed. In other words, "*nature is simultaneously real, collective, and discursive.*"[131] The decline of naturalism both gives way to a new conception of nature as artificially produced and compels individuals and collectives to hold various natures in tension.[132] The author proposes an anti-essentialist framework to identify nature regimes and argues for the inevitability of hybrid natures in the contemporary world.[133] Escobar suggests that the task of Political Ecology is to characterize the processes of articulation of the biophysical and the historical in order "*to suggest potential articulations realizable today and conducive to more just and sustainable social and ecological relations.*"[134]

Escobar has identified three main actors to illustrate the span of articulations of the biological and the historical. The first of these are the black and indigenous communities "*that for several centuries have been active in the creation of lifeworlds and landscapes of particular kinds*", which are, in fact mostly unfamiliar to us, Western scholars.[135] The second actor is a capitalist at work, in this case on a plantation. This actor argues that he "*is making development happen*" providing jobs to local populations. The third actor is a biodiversity prospector, interested in genes. Escobar suggests that each of these actors "*embod[ies] significantly different regimes of articulation of the historical and the biological*". He refers to these regimes as "organic nature", "capitalist nature" and "techno-nature".[136] However, he also states that these actors hold various landscapes in tension.

I now briefly introduce how Escobar characterises two of these nature regimes because they are particularly pertinent to this dissertation. On one hand, the capitalist nature results from a particular phase of history he refers to as patriarchal capitalist modernity. Escobar argues that capitalist modernity required the development of governmentality based on expert scientific knowledge, with an increasing appropriation of daily life by administrative apparatuses of the state.[137] Capitalism provoked the emergence of "society" from "nature". This separation resulted in the ensemble of social institutions "*which regulate the exchange of commodities, including the nature(s) produced by humans.*"[138] In short: capitalist nature is uniform, legible, manageable and harvestable. It is becoming an obstacle for the capital accumulation itself (see O'Connor "second contradiction of capitalism" and Harvey's "accumulation by dispossession").

On the other hand, the defining feature of the organic nature regime is that nature and society are not separated ontologically.[139] Kottak and Descola, along with others, have been showing that rural and indigenous communities "*signify and use their natural environments in very particular ways.*"[140] An important feature of these cultural models of nature is that they do not rely on a nature-society dichotomy. In fact, local non-Western models are often based on the links between the biophysical, human and supernatural domains.[141] According to Escobar, this continuity "*is culturally established through rituals and practices and embedded in social relations different from capitalist or moderns ones.*"[142]

Escobar suggests that the "*anthropology of local knowledge*" is the best suited discipline for the study of organic regimes. As evidenced by the important contributions of the "practice oriented" view of local knowledge, as proposed by Ingold and others, to the understanding of organic nature regimes. The cultural models of nature as ensembles of meaning and uses are based on historical, linguistic and cultural processes. He argues that its analysis from an anthropological Political Ecology approach, should be situated in the larger contexts of power and articulation with other nature regimes and global forces.[143]

Escobar mentions social movements from the rain forest that emphasize the fundamental rights to: territory, identity, political autonomy and an idiosyncratic view of development or economy. Thus, reflecting that their struggle is cultural, political and ecological.[144] As pointed by Martinez Alier these local groups necessarily engage in market exchanges while resisting a purely capitalistic valorisation of nature.[145] Escobar suggest that the hybridization of nature regimes "*might constitute for these groups an attempt to incorporate multiple constructions of nature in order to negotiate with trans-local forces while maintaining a modicum of autonomy and cultural cohesion.*"[146] This hybridization, he continues, "*could also shed light on the manifold economic forms (…) that are at play or being created in peasant and rainforest context.*"[147]

1.4.3 The perception of the environment and the "dwelling perspective"

For more than a decade, Ingold has insisted on the profoundly relational character of reality, which Political Ecology approaches from a neo-realist perspective. Besides phenomenology, his work is inspired by the ethnography of non-Western groups. According to his "dwelling perspective", life happens in the engagement with the world in which we dwell.[148] Prior to any objectification "*we perceive the world because we act in it, and we similarly discover meaningful objects in the environment by moving about in it.*"[149] This is how Ingold tries to overcome the essentialist/constructivist debate: as things are the result of a co-construction process they are neither "culturally constructed" nor "naturally" given. Following this reasoning, the perception of the environment as a mode of engagement with the world and knowledge of the world is obtained through a process of "enskilment" developed through the active encounter with things.[150]

Ingold's contribution has concrete implication for a better understanding of the Angaité relation with their territory, in general, and their livelihood strategies, in particular. He suggests that the world of hunter-gatherers "*is constituted through their engagement with it, in the course of everyday, subsistence-related practices.*"[151] As opposed to the culturalist approaches, Ingold highlights that the sphere of human engagement with the environment, in this case constituted by practical livelihood activities, is embedded in the sphere within which humans are constituted as social beings. He offers an alternative to perceptual relativism, the conventional anthropological claim that different cultures process the same data of experience in terms of alternative framework of belief or representational schemata.[152] Ingold argues that such perceptual relativism reinforces Western naturalism and shows that both claims are based on the division between humanity and nature, on one hand, and the division between "indigenous people" and "enlightened Westerners", on the other.[153]

Ingold suggest that both hunting and storytelling "*are in the first place, ways of dwelling.*"[154] Therefore, countering anthropological approaches to human environmental relations that separate material subsistence and cosmological construction of the environment. Comparing different ontologies of people considered hunters and gatherers, he states that *"there is only one plane, in which humans engage, as whole organism-persons, with the component of the environment, in the activities of procurement.*"[155] This leads him to make an important distinction: in *hunter-gatherer ontology* there is no nature/society division.[156] This contrasts with our (Western) understanding of nature as a "natural environment": "*we imagine ourselves to be beyond the world, and there is a position to intervene in its processes.*"[157] Ingold concludes that conventional anthropological approaches tended to exclude the forest and the hunter-gatherer's interactions within it, from the domain of the social relations.[158]

1.4.4 Cosmographies

Paul Little's cosmography concept is central to my analysis of the Angaité responses. The author is associated with the second generation of political ecologists. However, I argue that this concept remains useful and can be updated in order to insert it in current political ontology and anthropological debates. In his book *Amazonia*, Little uses this notion to refer to the *"collective, historically contingent identities, ideologies and environmental knowledge systems developed by a social group to establish and maintain a human territory."*[159] For Little, a territory denotes the "*appropriation of a specific portion of a region by a social group according to their respective cosmographical tenets.*"[160] Little defines territoriality, when he mentions indigenous cosmographies, as a cosmographical framework for human occupation that unites use patterns with cosmological principles.[161] Following the work of Wilbert and in a rather simplistic way, Little argues that indigenous myths serve as an ethnoecological blueprint of subsistence and residence.

Historically, cosmographies can overlap in time though one proceeds another. Cosmographies can also spatially overlap, with one group territoriality placed over part or all the territoriality of another group[162]. Little explores the historicity of an indigenous cosmography by researching those that influenced it. He concludes that the repeated and long superimposition of distinct cosmographies (mission, development, and conservation cosmographies) "*radically changed indigenous peoples and their cosmographies.*"[163] The resulting new territories have "*hybrid characteristics that incorporated elements of other cosmographies.*"[164]

In order to update this concept and insert it into current debates, I relate it to a set of contributions that focus on the embodied practices involved in the creation of places/territories. I aim to overcome the unquestioned Nature/Culture division underlying Little's work. His cosmography and territory concepts are associated with an established theoretical definition of places as socially constructed. This perspective was based on a dichotomy that distinguished space, an abstract domain without meaning, from place, a space that becomes meaningful through human experiences and perceptions.[165] As mentioned before, for Little, a territory refers to the "*appropriation of a specific portion of a region by a social group according to their respective cosmographical tenets.*"[166] This is, cosmographical ideas are attached to spaces to construct a place or a territory. The constructivist assumptions underlying both Little's work and the established theoretical perspective on place ignored the existence of multiple ontologies, some of which are not based on the nature-culture divide. Some scholars attempted to overcome the constructivist limitations by putting some emphasis on the practices that create place.[167] Gupta and Ferguson for example, focused on the process of place making through embodied practices that shape identities and enable resistances.[168] However, these attempts also reproduced the nature-culture divide by failing to overcome the space/place dichotomy and by limiting the practices, experiences or perception that make a space meaningful exclusively to the humans.[169] To overcome these limitations, and to include the agency of non-humans, scholars like Massey, Jones, Thrift, and Harrison tried transcend the realm of social (human) relationships by approaching places as connected and temporal projects.

Carolina Tytelman's practices of place production study, in the context of forest co-management in Nitassinan/Labrador, does not relate place to an abstract space, *"but instead to the web of relations and practices that conforms it."*[170] She defines place, influenced by Ingold's *"dwelling perspective"*, as *"an assemblage of enduring (but changing) patterns produced by the interlinked practices of people and non-human persons."*[171] She clarifies that place is not projected over a space, humans and non-humans are landscapes' elements and ideas regarding places can contribute to their production.[172] She calls the practices that contribute to the production of place, *"practices of place"*.[173] Based on this approach she argues that Nitassinan and Labrador *"are two distinctive patterned places, which are produced by the practices of heterogeneous assemblages of humans and non-humans."*[174] None of these two places are considered to be enclosed, complete and self-coherent. In fact, considering the colonial context and the unequal degrees of access to power, both places are entangled by sometimes contradictory practices. The effect of this entanglement is evidenced in the extent to which the production of Nitassinan is made difficult, or nearly impossible, by the production of Labrador. In arguing this, Tytelman relates the *"place practices"* with the activities considered by the Innu people as core of their identity and in which they can practice their fundamental values. It is through these practices, she states, that the Innu people *"re-enact their social relationships with human and non-human persons and contribute to the network of relationships that constitute Nitassinan."*[175]

Arturo Escobar's contribution on "nature regimes", Ingold "dwelling perspective" and on Tytelman approach to the "place practices" lead me to argue that Little's cosmography concept can be updated in order to overcome its abovementioned limitations. Only such an update makes it possible to use this concept in a third-generation Political Ecology framework. I will use cosmography to refer to the changing ideas and practices that contribute to the production of a particular place or territory; therefore, to the production of a particular understanding of the world. Following Tytelman, I conceive "La Patria territory", "La patria colony" and the cattle ranches as three *"distinctive patterned places, which are produced by the practices of heterogeneous assemblages of humans*[iv] *and non-humans."*[176]

1.4.5 Livelihoods and sharing in convivial societies

The most common definition of livelihoods refers to individuals, households or groups making a living, trying to meet their economic necessities, to cope with uncertainties and respond to new opportunities[177]. In the late 1990s livelihood studies were at the centre stage of development and interdisciplinary studies. Evidenced by the highly promoted "Sustainable Livelihood Framework," a tool developed by the British Department for International Development.[178] In the Sustainable Livelihood Framework, the livelihood strategies are the combination of activities that people choose to undertake in order to achieve their livelihood goals. The choice of strategies is understood as a dynamic process,

[iv] *Tytelman uses "non-humans" as defined by Hallowell (1960) based on the Ojibwa sense of personhood, where "person" contains subcategories (human-person, animal person, thunder person, etc).*

constrained by some structural factors. Though all five assets (human, social, natural, physical and financial capital) are fostered or constrained by the changing structures and processes of the wider society, the livelihood approach was criticised for neglecting the power relations.[179] Newer generations of livelihood studies tried to include and analyse the power relations involved in livelihood strategies.[180] However, in the last decade, research increasingly shifted away from this actor oriented livelihood perspective.[181]

Amber Huff mentions that "*complex social, ecological, and political processes affect how people make a living and whether or not that living is satisfying and sufficient to maintain health and wellbeing.*"[182] Ecological conditions, the goals and decisions of individuals and social institutions that mediate access to resources, determine whether or not an individual or collective can access and transform these elements into subsistence security.[183] As De Haan points out, a holistic approach to livelihoods, including cultural, social, economic and political dimensions is necessary for a better understanding of the complexity of livelihoods.[184] Actually, when Karl Polanyi wrote the first theoretical account on livelihoods, *The Livelihood of Man*, he wanted to develop a holistic and human centred economic science.[185] Another forerunner of modern livelihoods research is Vidal de la Blache, who used the concept of "*genres de vie*" (modes of life).[186]

Amerindian cosmologies counter our common Western understanding of economic production as a material relation with the environment. They suggest that people "*understand their relationship to the biotic environment not in terms of sustainable food production, but rather in terms of the sustainable production of fully socialized 'persons', through relations with spirits that cycle between positive and negative reciprocity.*"[187] In other words, the outcomes of livelihood strategies are, in other cultures, not strictly related to what we would commonly consider an economic strategy.[188] According to Rubenstein, this might be cultural relativism's most important contribution to Political Ecology as it calls attention to the contextual nature of values and choices. This was also problematized by Fischer in his book *Rain forest exchanges*[189], where he argued that indigenous societies of the Amazonia operate within different regimes of value and social structure. This is related to the classical formalist-substantivist debate in economic anthropology. Substantivist anthropologist, historians and economist demonstrated convincingly that in noncapitalist economies, wants and needs are not only socially constructed or encoded, but also that so-called economic activities in fact do have many other functions than maximizing utilities.[190]

Certain economic practices could have unexpected political and cultural implications. And this leads us to the particularities of convivial societies. Hannes Kalish points exactly to this issue when he explores the meaning that sharing had among the Enlhet[v] and that goes beyond material connotations:

> "*Sharing used to determine Enlhet communal life: people shared their food, shared their time, shared their whole life, and wanted everybody to be satisfied in the same manner, everybody would be included, in the same manner everyone would participate in a joy created together. The Enlhet refer to this manifestation of sharing as 'nengelaasekhammalhcoo'. Linguistically [...] means ´to do mutually´ and ´having a good relationship with one another.´*"[191]

As Villagra points out, this more comprehensive aspect of the meaning of sharing resonates with the concept of conviviality, understood as "*an aesthetic and peaceful joyful communal life created through the actions of autonomous individuals who emotionally and intellectually invest time and effort in the construction of such a life.*"[192]

Following a holistic approach, I analyse the Angaité livelihood strategies and the process of land use change. I argue that such an approach to livelihood contributes to understanding the everyday tensions and contradictions faced by ordinary members of La Patria. I consider that these points of friction and the related political conflicts are strongly linked to the redefinition of the "political" and the "social collective".

1.4.6 Adaptive resistance

In their introduction to *The land within* (2005) Pedro García Hierro and Alexandre Surrallés state that indigenous people secure a future for themselves by gradually integrating and adapting their institutions to the national legal systems, meanwhile strategically taking advantage of this process. However, they also argue that the price of this strategic adaptation is high: "*the original meanings of indigenous territorial visions and institutions have gradually become distorted. In fact today's titled lands are the result of a long history of*

[v] *From the same language family as the Angaité.*

minor and not so minor conflicts, along with agreements, renunciations, resignations and adaptations, all combining to make the situation confusing, even to the inhabitants themselves".[193] Bruce Albert states that the development frontier "*subjects the socio-symbolic coordinates of indigenous territories and the collective identities they sustain to disruptions that are as profound as those suffered by their systems of production*". It impels these groups towards the dynamic of *adaptive resistance,* which became a crucial dimension of their social and cultural reproduction[194] This scholar conceives *adaptive resistance* as a process in which indigenous peoples' identities and mechanisms for social and cultural reproduction come to depend "*as much on repertoires of legitimation imposed by developing states and advocacy organizations as on their own political-symbolic resources.*"[195]

This concept has been used by others scholar to explore the mechanisms and strategies developed by indigenous peoples "*to simultaneously adapt, re-enact, and resist the process of blending indigenous and Western social-political organizational structures and institutions through time.*"[196] I argue that the Angaité's responses to the changes considered in this dissertation are part of a long process of adaptive resistance to the dispossession of their territory.

1.4.7 Summing up

In this dissertation, I engage with the political ontology project to try to overcome the tendency to separate human-environmental relations. This dichotomy separates the cosmological construction or representation of the environment from the people's practical interaction with environmental resources in the context of subsistence activities.[197] In other words, I try to overcome the Western ontological dualism that takes "nature" and "culture" as distinct domains.[198]

I conceive the Angaité territorial struggle as an ontological conflict, a struggle that is about more than just land or resources. Analysing the process by which Angaité human-environmental relations change, from a political ontology perspective, does not imply rethinking how their "culture" relates to "nature". Rather, it implies reconsidering the assumed ontological division of the "human-environmental relations" concept itself.

I conceive human-environmental relations as constituted by interwoven discursive, material, social and cultural dimensions.[199] For descriptive and analytical purposes, I rely on the following concepts: ontology, cosmography, nature regimes, place/territory and livelihood strategies.

I use Blaser's definition of the ontology concept: "*the unstated but implicit understanding of how reality is constituted.*"[200] Ontologies, according to this scholar, are shaped by the practices and interactions between humans and non-humans.[201] Using this concept implies questioning the routine usage of the "culture" concept. It implies a notion of the world as a single reality "out there" as separate from the many cultural interpretations and representations of it.[202] Culture as a means to think differences among human groups is, for Blaser, an inadequate concept that takes for granted its own ontological status.[203] However, following Escobar, I argue that some notions of culture could and should be retrieved for strategic usage.[204]

I use Escobar's "nature regimes" concept, because it helps me understand how the Angaité articulate the nature-society dichotomy. A historical, rather than essential, model of nature links the biophysical, human and supernatural domains through rituals and practices. I argue that Angaité cosmography can be understood as an organic nature regime which is in growing tension with the cattle rancher's development cosmography, characterised as a capitalist nature regime. Based on Arturo Escobar's "nature regimes", Ingold "dwelling perspective" and Tytelman approach to "place practices" I use the cosmography concept to refer to the changing ideas and practices that contribute to the production of a particular place or territory; therefore, to the production of a particular understanding of the world.

1.5 Methodological considerations

I conceive this ethnography as storytelling, as a narrative of the struggle in which the Angaité are involved to respond to changes, and to sustain their worldings and life projects. Ethnography commonly refers to the integration of first-hand empirical investigation as well as theoretical and comparative interpretations of social organization and culture.[205] In terms of data collection, it usually involves the researcher's participation in people's daily lives, ways of watching, listening, asking questions and collecting documents, typically employing a relatively open-ended approach.[206] Initially, ethnography as social research was based on "naturalism" and thus understood as an alternative view to "positivism". However, naturalist ethnographies have been subjected to the abovementioned anti-realist and political critiques. They are forced to question the supposed neutrality or objectivity of all research. Since then, ethnographies are expected to be more reflexive. They acknowledge that the researchers' orientations are always shaped by their socio-historical location.[207]

In this ethnography, I engage with political ontology to try to overcome the modernist ontological commitments of the western academic knowledge practices. Paraphrasing Blaser, my ethnography as storytelling is not to explain or represent the Angaité and their transformation process. Instead, I narrate their responses to current changes in a tone determined by my particular way of relating to them and their embodied views of the world.[208] For many indigenous people storytelling has profound performative qualities. Stories do not refer to something "out there" but instead have a purpose and help to produce that of which they speak.[209] Approaching ethnography as storytelling serves a purpose: to contribute to the *pluriverse*, the way different worlds exist and thrive alongside each other.

I conducted ethnographic fieldwork research in Paraguay from May 2015 to April 2016. During this time, I lived in La Patria colony for ten periods of approximately fifteen days each. The periods in between, I remained in Asunción systematizing the field data and planning the next stages of data collection. My research design combined qualitative and quantitative methods as well as primary and secondary sources. A participatory mapping methodology was developed in the initial proposal. The research design had the characteristic "funnel structure" progressively focusing over its course.[210] The first three fieldwork periods were vital firstly to constructing a rapport with the leaders and inhabitants of different villages, secondly to delimit the research problem and lastly to familiarize myself with their use of the *Guaraní* - an indigenous language that I understand, speak and write.

1.5.1 Participatory mapping

After the third fieldwork period, the first phase of the participatory mapping methodology was developed to explore the socio-ecological setting of the region, the colony and the villages. This method was based on previous mapping experiences from the region.[vi] However, another important source was the "Sistema Jaguar".[211] The latter was initially developed to map the land use and occupation of indigenous people from the Argentinian Chaco and then adopted by the Argentinian *Instituto Nacional de Asuntos Indígenas.* My previous mapping experience with the *Mbya Guaraní* people from the

[vi] *See Carlos Salamanca and Rosario Espina, 2012, and Mac Chapin, Zachary Lamb and Bill Threlkeld, 2005*

eastern part of Paraguay was also relevant.[212] When I was developing the participatory mapping methodology I was conscious of the limitations and critiques of "indigenous mapping" experiences. In short, critiques point to the way in which concepts and categories used in modern cartography constrain the translation of indigenous knowledge and territorial use. Others argue that the use of GPS and other modern mapping technologies are inadequate for accurately conveying indigenous spatial understandings.[213] Considering these critiques and limitations led me to approach map-making as an opportunity to explore the current areas and resources used by the Angaité inside La Patria and, at the same time, to explore what they said about the tensions and conflicts related with this use. In other words, I was not trying to represent the "Angaité geography" nor was I particularly interested in the resulting maps; I was interested in the process.

The participatory mapping included: outlining the spatial pattern of livelihood activities, a basic description of ecological characteristics of the various land use areas and the spatial extension of land use patterns at the village and colony level. Modifications to this mapping methodology were supervised by Mr. Santiago, the leader of *Karoa'í* village, and then approved by the leaders of the 18 villages of La Patria.

The methodology used to produce the maps below required 1-3 days visits to each of the villages. Sketch maps drawn by someone from the village involved the active participation of leaders and villagers. My aim was to identify the lands and resources that villages used on a periodic basis for their livelihood activities inside the colony. However, the executed sketch maps combine my research interest and those of the particular villagers of each village. They generally included houses, roads, water systems, cultivated areas, pastures, fishing grounds, and hunting and gathering areas. Once finished with the sketch map, usually after one or two days of work, a transect walk was conducted. During the transect walk, each object or area identified in the sketch maps was geo-referenced using a GPS. Visiting each one of the villages for a couple of days created an opportunity to talk to leaders, their families and other villagers about different aspects of their land use, their history and the current tensions and conflicts. The first phase of participatory mapping 18 villages took three months of field work (see Table 1).

Table 1. List of villages of La Patria colony and month in which the participatory mapping was conducted.

No.	Village	Date
1	Tatarē	September 2015
2	Carpincho	September 2015
3	Primero de Mayo	September 2015
4	Puente Kaigue	October 2015
5	Tres Quebrachos	October 2015
6	Monte Kué	October 2015
7	Laguna Teja	October 2015
8	Colonia 24	October 2015
9	Seis de Marzo	November 2015
10	Karova Guasú	November 2015
11	Las Flores	November 2015
12	Urunde'y	November 2015
13	Paraíso	November 2015
14	La Paciencia	November 2015
15	Karoa'í	November 2015
16	La Leona	November 2015
17	Laguna Hū	November 2015
18	San Fernández	November 2015

During this participatory mapping period, my permanent camp was the *Karoa'í* village leader's house. From there I travelled by bicycle to the other villages for a couple of days, taking my mapping and camping equipment with me. I presented the sketch and final maps made using QGIS software to leaders from the 18 villages. Their reactions and feedback were recorded and registered in my field diary. I classified the results of this participatory mapping phase in 18 villages into 7 types:

Sketch maps: Describe the basic settlement information and give a general overview of current and past resource uses. The drawing of the sketch map itself presented an interesting approach to Angaité representation of space (see Figure 11).

Notes: Notes based on the observation and conversation during the sketch map drawing, the transect walk and the geo-referencing round. The data contains a general overview of livelihood strategies, types of resources used and settlement patterns.

GPS: GPS waypoints and tracks. Geo-referenced places and areas considered relevant to understand livelihood strategies as well as past and current resource use.

QGIS: QGIS table of attributes. Describes each object and its properties: house ownership, size of agricultural plots, agricultural production diversity, size of cattle paddocks, livestock variety and quantity, water reservoirs capacity, and historical data.

Pictures: Pictures of participatory mapping exercise (see pictures in the appendix).

Maps: Cartographic map based on geo-referenced (GPS) information. Combines the information obtained during the mapping exercise with digital satellite imagery (see appendix 5 and 6).

Census: population census made and used by leaders of each community. Most of these censuses only include the adults and are classified by "families".

Villagers usually agreed that the sketch maps were an apt representation of what they considered their territory within the colony. Figure 11, for example, represents the areas and resources used by the villagers of *Tatarē*. Mr. Diego, the leader, started the maps by carefully drawing the river and the roads; then, he identified the areas that are cyclically flooded; finally, he drew every house, their horticultural plots, the forest shared with other villages and the rest of the elements. Other villages and even Ceibo colony were included as references. He also included two fences considered by the villagers of *Tatarē* as historical limits with the neighbouring villages. It seems that path or routes provide the Angaité with an important reference in these maps. These borders and passages support the theory that some 'hunter-gatherers' perceive landscapes in terms of vectors as opposed to areas.[214]

The map-making experience was different in every village. In some cases, the map drawing exercise facilitated communication but in others it made it more difficult. This mainly depended on the personal characteristics of the leaders, their skills and experience. The use of maps, in general, was alien for all of them. However, some had acquired mapping skills with Andrew Leake 20 years ago. Instead of teaching them to draw "proper" sketch maps, maps that follow Western conventions, I focused on the conversations and discussions that emerged in the drawing and geo-referencing process. The accuracy of the final maps depends on the geo-references identified during the transect walks. Due to the request of some young Angaité, I trained them to use the GPS device. Once I finished the participatory mapping in each village I designed the land use maps using a particular software (QGIS). I gave one copy to each village leader and left another to be permanently displayed in the meeting hall in *La Leona* village. Soon leaders started to use this map when territorial issues were discussed. In this first phase I also explored the areas and resources used by the Angaité of La Patria outside their land property or colony. However, I relied on maps drawn on the ground during interviews, instead of asking interviewees to draw maps on paper.

A second participatory mapping phase was proposed and approved. The exercise involved 4 of La Patria colony's 18 villages: *Tataré, San Fernández, Tres Quebrachos* and *Karoa'í*. The main objective was to deepen understanding of the current livelihood strategies practiced by the selected villages through observation and description. The village selection criteria were population size, location/accessibility and the occurrence/non-occurrence of conflicts or tensions related to natural resource use. In terms of population size, two small villages (Tataré and *Tres Quebrachos*), a medium size village (*Karoa'í*), and a large village (*San Fernández*) were selected. In terms of location, one village with very difficult access (Tataré), one village that is periodically inaccessible (*Karoa'í*) and two villages located on the main road (*Tres Quebrachos* and *San Fernández*) were selected. Two villages (Tataré and *Karoa'í*) were selected for their low occurrence/non-occurrence of conflicts or tensions related to natural resource use. Also, two villages (*Tres Quebrachos* and *San Fernández*) were selected because they exemplified the opposite condition. This selection criteria enabled me to compare similar villages and at the same time show the great diversity of cases.

A specific combination of techniques was developed considering the objectives, time and resources. A "time allocation" study was planned but not applied because I considered it too intrusive. A series of visits, a couple of weeks long, was organised to avoid overburdening the leaders and their families in each village. The results obtained during this participatory mapping phase were classified into 4 types:

Semi-structured interviews: Applied to village leaders and their families.

Notes: Notes based on observations and conversations.

Census: Population census applied in 4 selected villages.

Survey: Material equipment survey.

1.5.2 Participant observation and interviews

The two participatory mapping phases were complemented with participant observation and interviews conducted inside and outside La Patria colony. Besides the abovementioned young local assistants and Mr. Santiago's invaluable support, no other assistants or translators were involved in this research.[vii] Particular attention was placed on the communal leaders' assembly and casual meetings. Most of them took place in *La Leona* village, where the communal meeting place of La Patria is located. A series of semi-structured interviews were conducted in several villages, involving leaders, their families and other inhabitants. These discussions explore their livelihood strategies as well as the local norms and institutions that regulate the access, use and management of natural resources. The best conversations took place in casual settings, such as the *tereré* (cold *mate*[viii]), drunk early in the morning or at any time of the day, and the long walks. To ensure the confidentiality and security of the people interviewed all names included in this dissertation were replaced with pseudonyms.

While living in the colony I was invited to participate in several special events. Some were audio or video recorded but all were described in the field diary. These special events included *choqueos*, leader's trips to other communities or places; evangelic religious mass; and special celebrations, such as La Patria colony anniversary celebration.

[vii] *Non-indigenous research assistants in Asunción helped me with the field notes transcriptions to Word.*
[viii] *Mate is a traditional South American caffeine-rich infused drink. It is prepared by steeping dried leaves of yerba mate (Ilex paraguariensis) in hot water and is served with a metal straw from a shared hollow calabash gourd known in Paraguay as guampa.*

Besides the proper participatory mapping exercise, I permanently observed the natural resource use and management. Specifically, I used the GPS waypoints and tracks to register the landscape changes, livelihood strategies and resource use and management observed within La Patria colony. The cattle ranches located in the region, some of them surrounding La Patria colony and other quite far away, were located using the GPS. Pictures were also taken systematically to register the changes observed along the fieldwork year.

All data collection methods above commented were complemented by a series of observations and interviews conducted outside La Patria colony, most of them in Asunción. Information from interviews was compared to data obtained in La Patria to clarify or corroborate facts. Besides these interviews, four events that took place during my fieldwork year were also observed: an *indigenists'* meeting at the Paraguayan National Parliament, an experts' reflection meeting about "*Indigenous leaders, challenges and future perspectives*" organised by a local NGO in Filadelfia city, open lectures related to indigenous territoriality and an internal group discussion organised by the NGO TierraViva about indigenous people and natural resources.

1.5.3 Limitations of research

My first visit to La Patria, years before I started my fieldwork, was in 2009 with Grupo Sunu, a Paraguayan NGO with more than a decade of work committed to Angaité cultural reaffirmation. Then, from 2010 to 2015, I got involved in Grupo Sunu as board member and as part time researcher. In 2010, I was also invited to join the board of TierraViva - an NGO with more than two decades of committed work related to indigenous people land restitution legal claims. Despite the fact that I conducted this research independently of Grupo Sunu and TierraViva, leaders of La Patria were aware of my relation to these organizations. This fact had some positive and negative outcomes, which were considered in order to reflexively analyse the field data obtained. The advantages provided included saving crucial time in rapport construction, being granted direct access to all the assembly meetings as well as leaders' informal discussions and being hosted in Mr. Santiago's house for the first few months. The disadvantages were that local families and some leaders found it difficult to see me and my work as independent from that of these NGOs. They tended to reproduce the same discourse that they would for a member of one of these organizations.

My gender also clearly influenced my fieldwork experience. I was working closely with leaders who are mainly male.

Based on Stephen Kidd's[215] experience, while living in La Patria I stayed with different families from distinct villages. My semi-intensive fieldwork strategy, 10 periods of approximately 15 days each - interrupted in January 2016 for 1 month due to dengue fever - gave me the chance to attain some understanding of the Angaité worlding and life project. However, a more permanent residence in the colony, for longer periods, would have probably led to a deeper level of knowledge.

1.5.4 Analysis of data

The analysis of data started in the pre-fieldwork phase when the research proposal was developed, continued during the data collection and ended with the systematization, manipulation and coding of the collected data. During each stage of this process, an initial idea emerged from the data. It helped me understand the information I had and needed to collect. As well as enabling me to link my ideas to those of others.[216] The periods I spent in Asunción between each visit to La Patria were crucial to processing and analysing the collected data before returning to the field to collect further data. As I was dealing with unstructured data, the post-fieldwork analysis involved firstly, the development of analytic categories to capture the relevant aspects of the data and secondly, the assignment of particular items of data to those categories.[217] Following Angrosino[218] and Gibbs[219] data analysis recommendations, I started with a descriptive data analysis. In this analysis, all data gathered was organised and classified in categories. Categories were not considered permanent, and were changed along the open-coding process. Significant patterns were identified in this stage. Afterwards, the identified patterns were contrasted with the existing literature on the subject. My findings were related to other interpretations. Atlas-Ti, Citavi, and the open source QGIS software were used for the coding, literature review and map design respectively.

---

[1] Ernst Halbmayer and Elke Mader, 2004
[2] Tim Ingold, 2000
[3] Ernst Halbmayer and Elke Mader, 2004
[4] Hans A. Baer and M. Singer, 2014
[5] Ibid.

[6] Ernst Halbmayer and Elke Mader, 2004
[7] Arturo Escobar, 1999
[8] Mario Blaser and Arturo Escobar, 2016
[9] Emily Yeh and Joe Bryan, 2015
[10] Roderick P. Neumann, 2005
[11] Mario Blaser and Arturo Escobar, 2016: 165
[12] Eric Wolf, 1972: 202
[13] Mario Blaser and Arturo Escobar, 2016
[14] Steven L. Rubenstein, 2004
[15] Raymond L. Bryant and Sinéad Bailey, 1997
[16] Ibid., 28
[17] Mario Blaser and Arturo Escobar, 2016
[18] Ibid., 167
[19] Ibid.
[20] Arturo Escobar, 1996
[21] Ibid.
[22] Roderick P. Neumann, 2005
[23] Mario Blaser and Arturo Escobar, 2016
[24] Steven L. Rubenstein, 2004
[25] Mario Blaser and Arturo Escobar, 2016: 167
[26] Tim Ingold, 2000
[27] Mario Blaser and Arturo Escobar, 2016
[28] Ibid.
[29] Ibid.
[30] Arturo Escobar, 2010
[31] Steven L. Rubenstein, 2004: 150
[32] Ibid.
[33] Mario Blaser and Arturo Escobar, 2016
[34] Ibid.
[35] Volker Von Bremen, 1987
[36] Ibid., 13
[37] Gastón Gordillo, 1993
[38] Mario Blaser, 2010: 89
[39] Gastón Gordillo, 1993: 73
[40] Ibid., 75
[41] Ibid., 81
[42] Ibid., 90
[43] Ibid., 91
[44] Ibid., 91
[45] Pedro García Hierro and Alexandre Surrallés, 2009

[46] Joe Bryan, 2009
[47] Ibid.
[48] Ibid.
[49] Tom Perreault, Gavin Bridge and James McCarthy, 2015
[50] ILO, 2003
[51] Emily Yeh and Joe Bryan, 2015: 533
[52] ILO, 2003
[53] Pedro García Hierro and Alexandre Surrallés, 2009
[54] United Nations, 2008: 10
[55] Ibid., 3
[56] OAS, 2010
[57] Ibid.
[58] Ibid., 1
[59] Rodrigo Villagra, 2010
[60] Andrew Leake, 1998
[61] Valentina Bonifacio, 2013b
[62] PNC, 2012: 24
[63] Ibid.
[64] Ibid.
[65] Nigel Rapport and J. Overing, 2003
[66] Ibid.
[67] Ibid.
[68] Emily Yeh and Joe Bryan, 2015
[69] Ibid., 532
[70] Ibid., 555
[71] Ibid.
[72] Ibid.
[73] Ibid.
[74] Joe Bryan, 2009
[75] Emily Yeh and Joe Bryan, 2015
[76] E. Cameron, S. de Leeuw and M. Greenwood, 2009
[77] Joe Bryan, 2009
[78] Steven L. Rubenstein, 2004
[79] E. Cameron, S. de Leeuw and M. Greenwood, 2009
[80] Ibid.
[81] Ibid.
[82] Ibid.
[83] Nigel Rapport and J. Overing, 2003
[84] Ibid.
[85] Emily Yeh and Joe Bryan, 2015

[86] Braun 2002 in Emily Yeh and Joe Bryan, 2015
[87] Brosius 1997 in Emily Yeh and Joe Bryan, 2015
[88] E. Cameron, S. de Leeuw and M. Greenwood, 2009
[89] Emily Yeh and Joe Bryan, 2015
[90] Shah 2010 in Emily Yeh and Joe Bryan, 2015
[91] PNC, 2012
[92] DGEEC, 2004
[93] Ibid.
[94] Emily Yeh and Joe Bryan, 2015
[95] Watts 2000 in Emily Yeh and Joe Bryan, 2015
[96] Emily Yeh and Joe Bryan, 2015
[97] Steven L. Rubenstein, 2004
[98] E. Cameron, S. de Leeuw and M. Greenwood, 2009
[99] Arturo Escobar, 1999
[100] E. Cameron, S. de Leeuw and M. Greenwood, 2009
[101] Mario Blaser, 2013: 547
[102] Ibid.
[103] Arturo Escobar, 2016: 21
[104] Ibid., 22
[105] Emily Yeh and Joe Bryan, 2015
[106] Mario Blaser, 2013: 562
[107] Mario Blaser, 2010: 3
[108] Ibid.
[109] Ibid., 2
[110] Mario Blaser, 2003
[111] Ibid.
[112] Ibid., 16
[113] Ibid.
[114] Ibid., 554
[115] Ibid., 565
[116] Ibid., 141
[117] Ibid., 558
[118] Mario Blaser, 2013:553
[119] Ibid., 547
[120] Mario Blaser, 2003: 33
[121] Ibid.
[122] Ibid., 34
[123] Ibid., 110
[124] Ibid., 132
[125] Ibid., 136

[126] Mario Blaser, 2010: 6
[127] Ibid.
[128] Mario Blaser, 2003: 551
[129] Ibid., 552
[130] Arturo Escobar, 1999: 1
[131] Ibid., 2
[132] Ibid.
[133] Ibid.
[134] Ibid., 4
[135] Ibid.
[136] Ibid., 5
[137] Ibid.
[138] Ibid., 7
[139] Ibid.
[140] Ibid., 7
[141] Ibid.
[142] Ibid., 8
[143] Ibid.
[144] Ibid.
[145] Joan Martínez Alier, 2004
[146] Arturo Escobar, 1999: 13
[147] Gudeman 1996 inArturo Escobar, 1999: 14
[148] Tim Ingold, 2000
[149] Ibid., 20
[150] Arturo Escobar, 2010
[151] Tim Ingold, 2000: 58
[152] Tim Ingold, 2000
[153] Ibid.
[154] Ibid., 57
[155] Ibid., 58
[156] Tim Ingold, 2000
[157] Ibid., 20
[158] Tim Ingold, 2000
[159] Paul E. Little, 2001: 5
[160] Ibid., 246
[161] Ibid., 14
[162] Ibid., 246
[163] Ibid., 79
[164] Ibid.,71
[165] Carolina Tytelman, 2016: 14

[166] Paul E. Little, 2001: 246
[167] Carolina Tytelman, 2016
[168] Gupta and Ferguson 1997
[169] Carolina Tytelman, 2016
[170] Carolina Tytelman, 2016: 17
[171] Ibid., 245
[172] Ibid.
[173] Ibid., 3
[174] Ibid., 18
[175] Ibid., 247
[176] Ibid., 18
[177] Marschke, M. J., and F. Berkes, 2006
[178] Leo de Haan, 2012
[179] B. Obrist, C. Pfeiffer and R. Henley, 2010
[180] Leo de Haan, 2012
[181] Ian Scoones, 2009
[182] Amber Huff, 2014: 165
[183] Ibid.
[184] Leo de Haan, 2012
[185] P. Sakdapolrak, 2014
[186] Ibid.
[187] Steven L. Rubenstein, 2004: 139
[188] Ibid.
[189] William H. Fisher, 2000
[190] Steven L. Rubenstein, 2004
[191] Kalish 2007 in R. Villagra, 2010: 51
[192] J. Overing and A. Passes, 2000
[193] Alexandre Surrallés and Pedro García Hierro, 2005
[194] Bruce Albert, 2005
[195] Ibid., 200
[196] S. Athayde and M. Schmink, 2014
[197] Tim Ingold, 2000
[198] Mario Blaser and Arturo Escobar, 2016
[199] Arturo Escobar, 1999
[200] Mario Blaser, 2003: 37
[201] Mario Blaser, 2010: 3
[202] Burman's review of Mario Blaser, 2013
[203] Mario Blaser, 2013: 550
[204] Escobar's review of Mario Blaser, 2013
[205] Martyn Hammersley and Paul Atkinson, 2007: 14

[206] Ibid., 16
[207] Ibid., 15
[208] Mario Blaser, 2010: 25
[209] Ibid., XV
[210] Martyn Hammersley and Paul Atkinson, 2007: 160
[211] INAI, 2009
[212] Carolina Rodriguez and Marcos Glauser, 2014
[213] Joe Bryan, 2009
[214] Ingold, 1986; Kidd, 1993; Villagra, 1998 cited by Andrew Leake, 1998
[215] Stephen W. Kidd, 2000b
[216] Martyn Hammersley and Paul Atkinson, 2007: 159
[217] Ibid., 174
[218] Michael V. Angrosino, 2012
[219] Graham Gibbs, 2007

2 Contextualization

2.1 The Gran Chaco and the Angaité

2.1.1 The Gran Chaco

Covering some one million square kilometres in size, the Chaco is geographically a central South American plain that encompasses portions of western Paraguay, eastern Bolivia and northern Argentina.[1] Its dry plains are considered the second largest biome in South America, after the Amazon.[2] The Gran Chaco region is delimited by the Paraná river and Paraguay river to the East, the Andean foothills to the West, the *Chiquitos* lowland and the Matto Grosso plateau to the North and by the Salado River to the South.[3] Average rainfall oscillates around 800 mm/yr.[4] Based on the pluviometry gradient the Gran Chaco is subdivided into three sub regions from East to West: The Humid Chaco, the Central Chaco and the dry Chaco. It is composed by a mosaic of ecosystems, predominantly the subtropical dry forest (to the West) and the sub-humid (to the East), adapted to a clear rain seasonality.[5]

In Paraguay, Argentina and Bolivia the Chaco represents pockets of forest coming under the increasing stress of demographic expansion and economic exploitation. In all of these countries the Chaco is also an ethnographic frontier, still open after centuries of contact.[6] According to scholarship on the Chaco region, its very existence as a distinct and important ecological and ethnographic area is seemingly overshadowed by its two neighbouring giants: the Andes and the Amazon.

2.1.2 Indigenous peoples of the Chaco and their territories

Some twenty indigenous groups, more than 250,000 indigenous people live in the Gran Chaco. According to traditional ethnolinguistic categorization these are classified into 6 main linguistic groups: Mataco-maká, Guaycurú, Lule-Vilela, Lengua-Maskoi or Enlhet-enenlhet, Zamuco and Tupí-Guaraní.[7]

These indigenous groups are bounded by a hunting-gathering tradition, common socio-political organizations and socio cosmological aspects.[8] According to the Chilean anthropologist, Nicolas Richard, the ethnologic specificity of the region pertains to the extreme malleability of its social formations, by which I mean its accelerated ethnogenesis processes.[9] Florencia Tola adds to this list, the generalised absence of matrimonial prescriptions, and the fluent ability to incorporate and transform external influences.[10]

Table 2. Chaquean indigenous linguistic groups

Linguistic group	**Indigenous groups**
Mataco-maká	Wichí-Mataco, Chorote, Nivaclé-Chulupí, Maká
Guaycurú	Toba, Toba-Pilagá, Pilagá, Mocoví, Mbayá-Caduveo
Lule-Vilela	Chunupí
Lengua-Maskoi or Enlhet-enenlhet	Enxet (Lengua sur), Enlhet (Lengua norte), Sanapaná, **Angaité**, Toba Maskoy o Enenlhet
Zamuco	Chamacoco-Ishir, Ayoreo
Tupí-Guaraní	Ava-Chiriguano, Chané, Tapiete, Isoseño-Guaraní, and Guaraní Occidental

Mainly based on historical sources, scholars agree that in pre-colonial times the *chaquean*[i] basic socio-organization unit was the band. They consisted of unions between many extended families whose members were interrelated by an alliance scheme. Bands are differentiated, in regional literature, from tribes. The latter are defined as less localised political units that group together various bands of a same dialectal variant or kinship.[11]

The conquest of the Gran Chaco, which started in the 19th century, had the following implications and repercussions: the use of terror and violence towards the indigenous population, the dispossession of their ancestral territories, the resettlement of indigenous population in marginal areas, the colonization by *criollo*[ii] population, the regional economic development based

[i] *Chaquean: inhabitant or object of the Chaco region. In Spanish: Chaqueño.*
[ii] *Crillo: non-indigenous Paraguayan or Argentinian population.*

mainly on indigenous slavery, the ideological prosecution of indigenous population and the conversion to Christianity as an assimilation mechanism. The different indigenous groups were constantly adjusting to these historical circumstances and generated particular configurations in order to achieve the perpetuation of their social specificity.[12]

Today's indigenous communities are disseminated in a vast territory occupied for over a century by ranchers, extractive industries, farmers, soybean enterprises, Mennonites colonies, NGOs, catholic and protestant missionaries and state institutions. These are just some of the non-indigenous actors and institutions that are present in the Gran Chaco.[13]

2.1.3 The Angaité in the Paraguayan Chaco

The Angaité are one of the six indigenous groups that integrate the *enlhet-enenlhet* language family.[iii] The latter are groups of languages that share some grammatical structures and vocabulary similitudes. In the case of the *enlhet-enenlhet* family, the close relation between its six members (Toba Maskoy, Enlhet Norte, Enxet Sur, Sanapaná, **Angaité**, Guaná) is easily recognised.[14] Hannes Kalish studied how some of these groups (including the Angaité) experienced a language loss process that held tragic consequences for the transmission of culture to new generations. However, this process is not new. Around 1967 Guaraní was already the main language some Angaité communicated in.[15] They currently speak "Paraguayan Guaraní", a language spoken by a 95% of the national population. Some people, mainly elders, still speak Angaité.

The Angaité, protagonists of this dissertation, are commonly and officially referred to using this term. They call themselves *enenlhet* and their language *ang-kayte pa'ayvoma.*[16] The Angaité as an ethnic group emerged from internal dynamics and subsequent colonial influence. Thus, Villagra suggests that identifying the pre-colonial groups which ended up merging under the ethnic label "Angaité" does not aid in reconstructing their identity.[17]

[iii] *In the 2012 Indigenous Population census the indigenous peoples that live in Paraguay were officially classified in 5 language families: Guaraní (Guaraní Occidental, Aché, Avá Guaraní, Mbya Guaraní, Paĩ Tavyterã, Guaraní Ñandeva),* ***Lengua-Maskoy or Enlhet-Enenlhet*** *(Toba Maskoy, Enlhet Norte, Enxet Sur, Sanapaná,* ***Angaité****, Guaná), Mataco Mataguayo (Nivaclé, Maká, Manjui), Zamuco (Ayoreo, Ybytoso, Tomárãho) and Guaicuru (Qom) DGEEC, 2014.*

Figure 2. Indigenous communities per ethnic groups in the Paraguayan Chaco. (DGEEC, 2004)

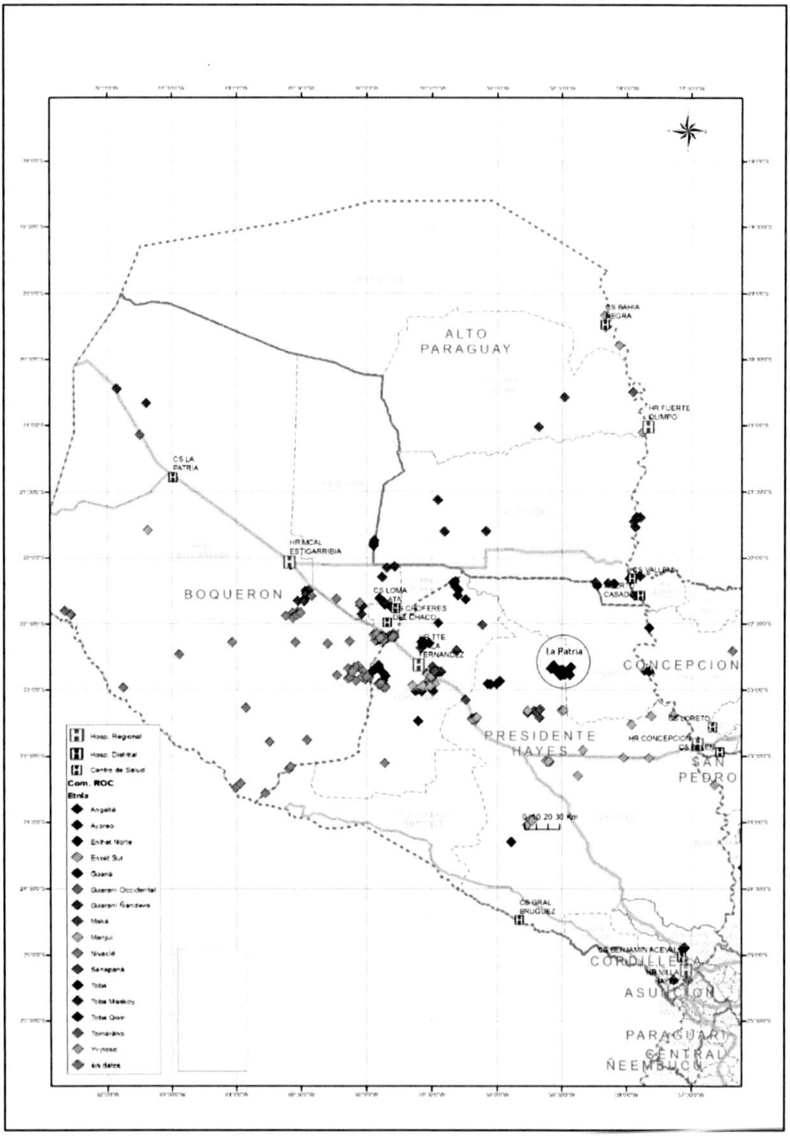

The 2012 Indigenous Population census provides information about the Angaité. This census considers "indigenous" any person who declares him or herself a member of an indigenous group and a community, village/neighbourhood or families nucleus. Thus, evidencing a disregard for the use or loss of the original native language.[18] Out of the 113,254 indigenous persons that live in Paraguay, 5,992 identified themselves as Angaité. The Angaité are in the seventh position of the ranking.[iv]

Table 3. Location and estimated population of Angaité communities

Community	District	Villages	Estimated Angaité population
La Patria	Puerto Pinasco[v]	17	1963 (32%)
Ex-Cora´í	Puerto Pinasco	6	659 (10.9%)
Riacho San Carlos	Puerto Pinasco	3	820 (13.6%)
Gente Rory	Puerto Pinasco	1	189 (3.1%)
San Martín	Filadelfia	1	112 (1.8%)
Santo Domingo	Filadelfia	1	163 (2.7%)
Diez Leguas	Tte. 1° Irala Fernández	6	1007 (16.8%)
Other[vi]			1079 (18%)
		Total	**5992**

Source: based on data from DGEEC 2014

This census provides us with varied yet relevant set of facts about the Angaité. Only 9.4% speak the Angaité language. Most of them live in rural areas. 71.9% have a Paraguayan identification document. 55% of the

[iv] *The biggest indigenous group is the Mbya guaraní (20.546 persons), followed by the Avá Guaraní (17.921) and the Paĩ Tavytarã (15.494), the three of them from the Eastern region. In the 3rd, 4th and 5th place are three groups from the Chaco region: the Nivaclé (14.768), the Enhlet Norte (8.167) and the Enxet Sur (7.284)*

[v] *Some villages of La Patria are actually located in Villa Hayes district. Every Governmental institution has his own classification. This has, indeed, caused great confusion.*

[vi] *These Angaité families live in two small settlements (Calería Itakuá, 13 persons), (Limpio, 10 persons) either in different indigenous communities which major population belong to a different indigenous peoples group (Makxawaya, La Herencia, Buena Vista, Karaja Vuelta, Comunidad 96, El Estribo, La Armonía, Nich'a Toyisy, Nueva Promesa, Casanillo, Pozo Amarillo, Laguna Pato, Cerro Poty, Redención, Limpio, Benjamín Aceval, Uj'elhavos, Colonia 5, Yalve Sanga, Pesempo'o, Livio Farinha, Riacho Mosquito, Castilla, San Isidro Km. 39, Maria Auxiliadora Km. 40, Machete Vaina y Boquerón-Cue)*

population older than 10 years old reads and writes and 51% of the population from 5 to 19 years old attends school. 19.6% used a cell phone in the last 6 months prior to the census and only 0.3% used a computer in the same period.

2.1.4 Angaité traditional livelihood strategies and land use patterns

It has been suggested that the ancestors of the *enlhet-enenlhet* lived in the Chaco inland, also called Chaco Boreal. This explains why historical documents mention little or nothing about them, if compared to groups that lived closer to the first European incursions along the Paraguay and Pilcomayo rivers.[19] According to historical documents (see Azara 1809; Aguirre 1793) the *enlhet-enenlhet*, usually referred to as *lengua* or *machicui*, originally lived closer to the northern bank of the Pilcomayo River. They moved closer to the Paraguay River after the mbayá groups disappeared.[20] According to Villagra, the *enlhet-enenlhet* people took the initiative to encounter the whites.[21] They travelled from distant western areas to the Paraguay River where they built camps to trade with merchant ships and towns. Branislava Susnik reported contact with the Enxet opposite the city of Concepción as early as 1841 and Cominges mentioned the Angaité around 1880.[22]

Leake generally described the native population of the Paraguayan Chaco traditional subsistence strategies and social organization. According to him, these societies were based on an egalitarian value system. The population lived in small-nomadic bands. Their subsistence economy was based on the sharing of food, the free access to natural resources and the autonomy as well as mobility of the individual as key social practices. According to Leake, each band was associated with an area of land sufficiently large for gathering wild fruits, collecting honey, fishing and hunting for their survival. Besides this, indigenous people also had domestic animals and kept small gardens. In short, their subsistence strategy was founded on a diverse and adjustable range of subsistence activities. In his 1963 contribution to Julian Steward's "Handbook of South American Indians", Alfred Metráux also referred to the subsistence of these groups: *"The seasonal yield of certain plant species produces a varied diet, and the irregular distribution of certain plants and several animal species induces a limited nomadism, which, however, does not involve the migration of large bands, but rather the dispersal of small family groups, which scatter in order to procure their livelihood. The social and ceremonial life is deeply affected by the momentary abundance of a particular food."*[23]

According to John Renshaw, indigenous subsistence strategies were based on immediate benefit return and the satisfaction of immediate needs.[24] This was also mentioned by Grubb, one of the first missionaries, who observed that *"[The Indian] never saves anything against the day of adversity (…). The clever hunter can bring in abundance of game, but he sees no need to unduly fatiguing himself, and is content with supplying his own immediate needs and the wants of those whom he desires to help."*[25]

Traditional subsistence strategies and contemporary livelihood strategies can be better understood alongside the indigenous perspective of a desirable social practice. Villagra suggests that the Angaité philosophy of desirable social practice is related to their concept of "living well" (in Angaité *takmelak yahekamaha,* what literally means "*to have a good way of being*").[26] Villagra follows Overing and Passes, among others, to suggest that this aspect is shared not only by other *enlhet-enenlhet* but also by other indigenous peoples of the Chaco and Amazonia.[27] For all these groups, "*to live well means a peaceful life among kin and co-residents that encompasses a wide range of concepts and practices of positive material and affective value, such as amity, love, respect, commensality and sharing. Living well requires constant commitment on the part of Ego, a requirement whose fulfilment can never be achieved beyond certain degree."*[28] Though disrupted by contemporary conditions under non-indigenous domination, this ideal is still present.

Leake estimates that these bands of 30-60 individuals used 7,500-15,000 hectares.[29] Based on his ethnographic records Villagra indicates that the *Koeteves* (Angaité subgroup) lived approximately 70-80 kilometres west of the Paraguay River and the *Kovalhok* (Angaité subgroup), whose territory reached the banks of the Paraguay River, were their eastern neighbours (see Figure 3).[30] He also mentions that these territories overlapped with other Angaité or Sanapaná groups.

As suggested above, mobility was a key aspect of the traditional subsistence strategies for the hunter-gatherer societies of the Chaco. Renshaw and Leake suggest that moving to an area where food was available was fundamental for the indigenous people to cope with the seasonal food availability in the Chaco.[31] This modality also made use of resources sustainable. People avoided depleting local game stock by periodically moving to new sites.[32] The landscape conditioned the spatial pattern of land-use because local bands were associated with particular subsistence zones, including permanent or semi-permanent water and food resources.[33] Susnik indicates a sense of association with "*vital areas*".[34] Apparently, private property existed in the form of cultivated garden plots[35], but according to Villagra, "*the concept of property in the juridical sense was alien…*"[36]

Figure 3. Angaité traditional territories (in 1900), La Patria territory (in 2015) and La Patria colony

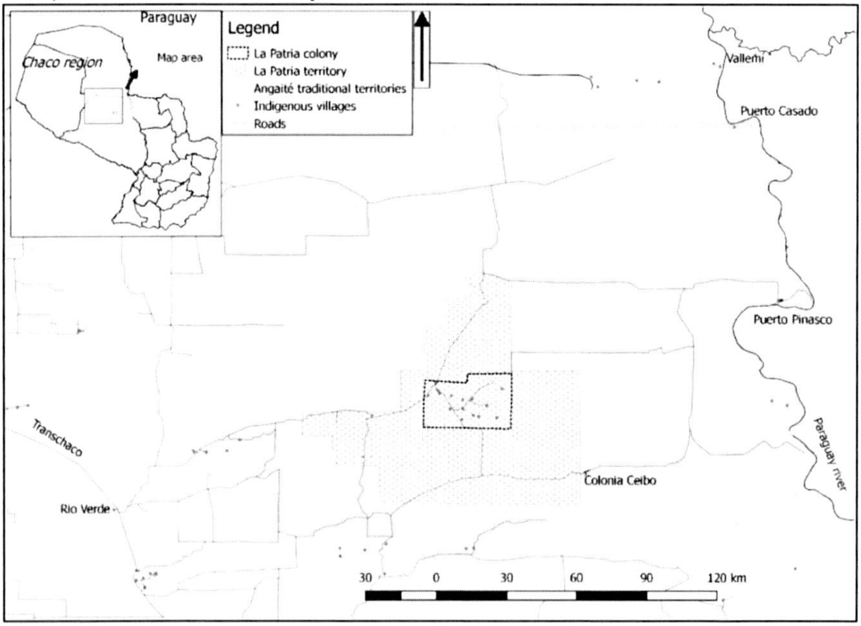

The Angaité traditional territories polygon is based on Villagra's map (see 2010, 132)

2.1.5 Historic process of territorial dispossession

The colonisation, the subsequent privatisation of land and the erection of fences, at the end of 19th century, restricted and eventually impeded residential mobility.[37] The influence of the white man dates back to Portuguese and Spanish explorations seeking a route to Peru in the 16th century.[vii] However, the Chaco still remained an indigenous territory

[vii] *"Asunción -the capital- was founded in 1537 and became the main centre of Spanish colonialism until the development of Buenos Aires in the 1580s. Spanish domination expanded swiftly within what is today the*

populated by diverse indigenous groups after the end of the Spanish colonial period (1537-1811).[38]

The Chaco was appropriated by the Paraguayan nation-state by a decree from 1825 issued by the Dictator Francia (ruling 1811-1840) and subsequently reinforced by President Carlos Antonio López (ruling 1842-1862).[39] The indigenous population of the Chaco was neither officially compensated nor aware that their land was sold by the state.[40] After the Triple Alliance War (1865-1870), with the "Public Land Sale Law" of 1885, the Government offered 98% of the physical national territory to mainly foreign investors in order to repay its war debt.[41] This sale of this concentration of land condemned the indigenous people to land dispossession.[42]

The main colonisation agents were the tannin companies, landowners (ranchers), missionaries, Mennonites and militaries. Their speed of incursion from East to West depended on the size of their properties, resources and their interests.[43] The South American Missionary Society (SAMS) established the *Makxawaya* mission to the South of Angaité territory at the turn of the century. They played a controversial role in "pacifying"[viii] the indigenous people and facilitating access routes into the region for white settlers.[44] The first market oriented activities in the Chaco included the extraction of *quebracho* trees (*Schinopsis balansae)* for the production of tannin. Followed, a few decades later, by livestock farming with extensive cattle ranches.[45] In 1887, the Carlos Casado Company established its headquarters to the north of Angaité Territory. This tannin company advanced into the northern Angaité, Sanapaná, Guaná, Yshiro and Enlhet territory with a 160 kilometres railway line.[46]

The heart of the Angaité territory remained untouched until the early 1910s.[47] The *Puerto Pinasco* town/factory was founded in 1907 by the Rosarina Company in the West bank of the Paraguay river and was later bought by the International Products Corporation (IPC). This company bought around 1,000,000 hectares, intensifying the hinterland occupation for timber

eastern half of Paraguay, where colonists mixed with the Guaraní tribe whose language is today spoken by the majority of Paraguayans. Expansion into the Chaco met with violent resistance" Leake, 1998.

[viii] *"Amongst the Angaité, the Anglicans founded two missions: Laguna Rey, which was situated 70 kilometres north of Makxawaya and lasted only a few months, from October 1928 till January 1929 (...), and Campo Flores mission – known in Angaité as Maskoykaha – which was established in 1930, 60 kilometres north-west of Makxawaya, and abandoned in 1946." Villagra 2010:93, Stephen W. Kidd, 1992; Andrew Leake, 1998*

exploitation and cattle-raising activities.[48] By 1931-1932, IPC had three well established ranches: *Kilometro 80* (later known as *Tuparendá*), *Riacho Gonzalez* and *Laguna Tigre*. All of them were situated beside lagoons or streams.[49] There was an 80 kilometres railway connecting *Tuparendá* and *Puerto Pinasco* and a 50 kilometres railway to the south west (today *Colonia Ceibo*). Villagra mentions that until 1940, when the control over land was secured, the Angaité approach to newcomers "*swung from cattle-hunting and retaliations to trading and bartering*."[50] The western limits of the IPC property remained unfenced until 1943. Parts of the Angaité territory remained open. In 1943 the company fenced off the area as reaction to the 1940 Agrarian Law.[51]

In Combès foreword to *Capitalismo en las selvas: Enclaves industriales en el Chaco y Amazonía Indígena (1850-1950),* she argues that anthropologist tend to portray the challenges of new relations between indigenous peoples and economic enclaves of mid-19th century from a bleak perspective. They focus on the enslavement conditions of indigenous work, the debt-trap, alcoholism and on acculturation.[52] The Angaité and Sanapaná were reported in several occasions to be among the most acculturated indigenous peoples in the Chaco, *"losing their language, traditions and even the capacity of production."*[53] Nevertheless, Villagra reveals in his doctoral dissertation that Angaité myths and narratives claim that they were deprived of their land by explorers, cowboys and soldiers, not by those who "legally" sold it.[54] Villagra discovered through myth analysis that the Angaité "*assume the blame for their land dispossession*". This assessment counters the common assumption that indigenous peoples consider themselves the victims. From this perspective, "*the breach of reciprocity with the Paraguayans*[ix] *provoked by the Angaité's failure to recognise an original exchange of food for land is central to the explanation of historical change and causation.*"[55] He also warns that this historical retrospective should not be understood as an alienated interpretation of history which justifies present domination. Rather, it is the way the Angaité become protagonists of their own history.

Employment in the tannin industry in *Puerto Casado* and *Puerto Pinasco* and cattle ranches offered the Angaité (among other indigenous peoples) a complementary source of income until the industry crashed. Indigenous

[ix] *"In general for the indigenous peoples, the Paraguayans belong to a specific ethnic category that applies to the members of the Paraguayan nation state and tradition, descent of the antique Guaraní-Spanish mixture." Villagra, 2010*

people represented 40% of the factory workers and the majority of the workforce on the cattle ranches in *Puerto Casado.*[56] When the tannin industry collapsed the ranches were subdivided into smaller units or *estancias.*[57] Though land was privatized, most indigenous groups continued to live within their traditional territories until land was progressively subdivided with fences to restrict free access to their traditional territories. Kidd mentions that the cyclical use of different resource areas was further inhibited because ranches were established next to waterholes traditionally used as camp sites.[58] Since then, the Angaité were forced to live in fixed locations determined by the ranch owners and exploited as cheap source of labour.[59] By the end of the 1970s, some families moved to the *Makxawaya* Anglican Mission or the Mennonite colonies and the rest remained living in/alongside cattle ranches.

It is important to mention that although the importance of wage labour increased during the tannin boom, subsistence activities continued to provide an important part of the diet.[60] This aspect was also underlined by one indigenous leader of La Patria who was born and raised on *estancia* 14 de Mayo:

> *"I think the ranchers knew how to manage the ranch. Every month they provided food, yerba mate, manioc flour, locro, maize, only that, in packages of 3 kilograms. There was also rice, pasta and galleta (bread) but that was for sale and to be discounted of the salary. (…) Then the food was over, but they slaughter their cattle every week. They had plenty. And if there was nothing, the workers would not complain because there were plenty of wild animals. In the afternoons, at this time, they [the indigenous workers] stopped working and went to hunt on horses. Horses from the company. Sometimes they brought ostrich eggs, late at night, around eight, in the darkness. (…) There were many ostriches, and people knew when they lay eggs, in which month. Others gathered wild honey and used their sheep skin saddle to bring the honey and the honeycomb. They didn't have where to put it, to avoid leaking… (…) In those years there was no paper, and no plastic and those kind of stuff".*[×]

× *"Che ahecha omba'apo porã la estanciero kuéra, ha cada mes ome'ẽ la provista, tres kílonte voi, ome'ẽ ka'a, fariña, lókro, avati, pévante. upéva la libre. Upéi katu arroz, fideo, galleta, upéva oho ikuéntape. (…) Upérõ provista opáma, pero la so'o semanal okarnea. Oreko hikuái. Ha ndaipórirõ mba'eve ndoprotestái la omba'apóva porque heta la vícho ka'aguy oĩ. Ka'arukue kóva óra ya opoíma hikuái ha omariska hikuái,*

2.1.6 Process of territorial restoration

In 1975, 150 years after Dictator Francia's decree (a formal dispossession of indigenous people), the Paraguayan law established procedures enabling indigenous peoples to submit land claims to state institutions.[61] However, due to the landowners' influence on the government and the fact that Paraguay was ruled by a longstanding military dictatorship[xi] at the time, legal claims were reduced to small extensions.[62] In those years, the declared aim of the Paraguayan government towards the indigenous peoples was to integrate them into national society by grouping the population into organised settlements. This integration process was mainly implemented by church-based organisations.[63] Already after the Chaco War (1932-1935), General Belaieff claimed that: "… *the majority of the Indians are fully aware of the necessity of readjusting their lives. Some of the Lengua, Angaité and Ashluslay (Chulupí) are already accustomed to farming, which they learned long ago in the missions ... The tribes are eager to participate in the progress of the country."*[64] There was a general assumption that resettlement into agricultural colonies was the only viable option for the indigenous peoples, resettlement was conducted in a similar pattern in North America.[65]

Though there was legal recognition of the indigenous peoples' right to access their traditional hunting grounds, the landowners often denied permission.[66]

In 1975, *"the Enxet indians asked the Anglican church for help in acquiring lands of their own."*[67] An anthropological evaluation of an Anglican project, conducted by Grünberg et al. a few years later, concluded that there was *"no alternative but helping them to settle on their own land."*[68] This was the origin of *La Herencia* project (funded by European donors) that bought the land and created La Patria colony.[xii]

kavaju ári. Empresa kavaju. Oĩ sapy'ánte alguno ogueru ñandu rupi'a, tárdema oguahẽ, la ocho umíva, pytũma oguahẽ. (…) Heta la ñandu, ha'ekuéra oikuaa ombo'a jave. Mba'e méspa. Alguno oho ombyai la rréina. Ha omoĩ ovechapire la hykuere, ipanal. Ndorekói la omoĩ va'erã. Pono otyky. (…). Anteve ndaipóri la kuatia, húle umíva ndaipóri." Mr. Santiago, February 2016

[xi] *Alfredo Stroessner's dictatorship was the longest in South America (1954-1989). The Colorado political party continued ruling the country until 2008. In 2013 this political party won the national elections and is currently ruling again.*

[xii] *La Herencia purchased Sombrero Pirí (9,500 hectares) in 1980 for the Enxet; La Patria (22,520 hectares) for the Angaité and Sanapaná in 1983; and El Estribo (9,375 hectares) for the Enxet.*

La Herencia project was based on the Mennonites resettlement model[xiii] and aimed to acquire 30,000 hectares of land to support a total of 300 families.[69] According to Leake: "*when these estimates were made, it was recognised that hunting and gathering still provided the Indians with an important part of their subsistence requirements (...) It was, however, deemed unrealistic for both financial and political reasons to even contemplate acquiring lands which might be large enough to allow the people to continue with these traditional activities.*"[70] Kidd mentions that when he started working on the project he was *"struck not just by their great poverty but, even more, by the fact that everyone was poor."*[71] From his point of view, the programmes of land acquisition undertaken by Anglicans, Mennonites and Franciscans between 1979 and 1989 legitimated the dramatic reduction in land size under the direct control of the Enxet (to only 4% of their former territory) and inhibited survival purely from traditional subsistence activities.[72]

Besides securing the land, the aim was to *"help them develop economies based on cash-crops, cattle and small businesses (...) encouraging them to make the transition from foraging to farming."*[xiv] Indigenous people did not respond as expected. In terms of production, few people engaged in agriculture. Hunting and fishing continued to provide an important source of food. Livestock raising was not widely adopted. In terms of remuneration, indigenous people didn't save or invest their income. The clan or extended family remained the strongest economic unit. Wage labour remained important to all resettled families, even where payment was very low, *"because they valued the chance of maintaining social links with other Indians on those ranches."*[73]

At the end of the 1980s, anthropologists Renshaw and Von Bremen sought to understand why the development programmes for indigenous people failed. For Renshaw the main reason was that "*planners have not understood the Indian's perceptions and value systems*", arguing that, despite the pressures and changes forced by the white man, they managed to "*maintain values and aspirations that are quite distinct from those of the national society.*"[74] He clarified that these values were not exactly the same as they were over a century ago but that "*they still conform to a vision of society, and of man's relation to the natural world, that is in many respects peculiar to the Indian societies of the Chaco.*"[75]

[xiii] *Villages units of 50 families with 100 hectares each were planned. "They would use 2 hectares for cotton, 1 hectare for castor oil seed, 1 hectare for horticulture and 96 for catlle pasture" Leake, 1998*

[xiv] *"...resettlement also entailed the Indians making the transition from a semi-nomadic hunter gatherer mode of subsistence to a sedentary existence based on agriculture." Leake, 1998*

It was in this context that Von Bremen developed his argument about the persistence of a hunter gatherer mentality, mentioned in the previous chapter. According to Blaser it had an enormous influence on the Paraguayan indigenist field. Von Bremen exposed this argument in a development projects evaluation report requested by German churches. This report became a controversial document as part of a regional debate on which mainly chaquean anthropologist were involved. Von Bremen's main argument was that the responses that the indigenous people of the Chaco gave to the new challenges, those that arose as consequence of the destruction of their traditional way of life, were based on the same logic that underlines the relations that these groups maintain with nature.[76] He argued that some conceptions and values of their historical hunter-gatherer way of life were still determining their conduct towards the development projects.[77]

2.1.7 Current challenges of indigenous peoples in the Paraguayan Chaco

Kalish argues that all indigenous groups from the Paraguayan Chaco still maintain a hunter-gatherer tradition. However, he also points out that the fragmentation into smaller groups, as result of the colonisation process, translated into different historical experiences. He suggests some parameters to understand present day differences. Some groups that have formal ownership over parts of their own lands while others are struggling to obtain this right or have not started the land claim process and do not own land. Some groups experienced a hostile contact with whites but in other cases contact with whites was not physically violent. Some have been intensively accompanied by white advisor for decades while others still live quite isolated. Besides some Ayoreo groups that still live in voluntary isolation, without any contact to external society, all the rest experienced centuries or decades of (Catholic and/or Evangelic) missionary work. These differences translate, according to Kalish, into radically different communal dynamics between communities that belong to the same indigenous group[78]. According to Blaser, the territoriality and responsibility towards land and nature that some indigenists attribute to indigenous people are actually rooted in this specific historical experiences of each group.[79]

In 2014 the indigenous population of the Paraguayan Chaco either owned or had legal rights to occupy a total of 839,313 hectares of land. In 2000, Kidd suggested that, although no ecological studies have been undertaken to determine the area needed by indigenous people, there is considerable evidence to affirm that the secured indigenous land were below the minimum required for indigenous livelihoods. Stahl suggested that every indigenous family would need at least two hundred hectares.[80] According to the Paraguayan legislation (Law 904/81), every indigenous family has the right to minimum 100 hectares in the Chaco region[xv] and 20 hectares in the Eastern region. This has been used by Oscar Ayala, a member of TierraViva, as a satisfaction indicator to measure the realization of this indigenous right.[81] According to his results the Chaco region is inhabited by 53,879 indigenous individuals, equivalent to 9796 families. The division of 839,313 hectares by this 9,796 families results in 85 hectares per family, not achieving the minimum of 100 hectares.[xvi] In 2014, the Angaité people owned a total of 46,114 hectares out of the approximately 1,000,000 hectares occupied and used before colonisation.[82]

According to the 2012 Indigenous Population Census, 27,6% of the indigenous communities in Paraguay do not own land and 36,9% of those who do are involved in irregular land leases.[83] Besides the ongoing effort of indigenous organisations and some NGOs to secure more land for present and future generations, indigenous peoples are also facing challenges related to the enjoyment of their territorial use rights and management of the lands already secured in the past decades.

A good example to understand the complexity of current challenges can be drawn from the conclusions that Leake gave in 2006 to a study requested by the Inter-American Court of Human Rights (IACHR), in relation to the *Sawhoyamaxa* Case.[xvii] The IACHR wanted to know the extent to which the 14,404 hectares reclaimed *"will be sufficient in size and quality to enable the indigenous people of Sawhoyamaxa to preserve and develop their particular way of life."*[84] In his answer to the IACHR Leake mentioned that "*contemporary indigenous livelihood strategies (…) tend to draw on resources and employment opportunities over broad regions*" and that *"land should be looked at in terms of how it enables indigenous*

[xv] *According to Leake (1998) the 100 hectares were established based on the Mennonites resetllement programs of mostly northern Enxet in agricultural colonies, and became accepted as a standard across the Chaco and eventually was enshrined in the national legislation.*

[xvi] *Ayala also found that the situation is worst in the Eastern region, where only 163,956 hectares were secured for indigenous communities but there are 10,274 families. According to these estimations, in this region every family has access to only 15,9 hectares, way under the legal standard.*

[xvii] *This case concerns the state's failure to ensure the ancestral property right of the Sawhoyamaxa community and its members.*

households to strengthen and develop their current livelihood strategies in accordance to their own priorities."[85] He also mentions that all imposition of settlement and land-use restrictions by external authorities should be understood as a violation of indigenous people's sovereignty and self-determination.[86] In his conclusion, Leake states: "*The land is suitable in this regard because it will provide a secure space within Sawhoyamaxa ancestral territory and thereby empower them to take their first tentative steps towards a future that they can build for themselves.*"[87] It is clear that the land secured and titled as "indigenous communities" in Paraguay is not enough. Therefore, livelihood activities include areas and resources located beyond communal land limits.

Surrallés and García Hierro share a very similar view: "*ownership of the land implies full enjoyment of a modus vivendi the autonomous continuity and evolution of which has been threatened for a long time.*"[88] However, they also warn that there are several factors that challenge the revival of their own production models and latent social networks. One of these factors is the current superimposition of other spaces on indigenous territorial space. Distinct identities and symbolic ties have been built and specific social relationships have been woven into each. The risk, they say, is that indigenous people can lose control over their spatial decisions and that their territories can sometimes turn into "object spaces" ready to be exploited by others.[89]

Indigenous Organizations, like *Nengvaanemkeskama Nempayvaam Enlhet*, identify the indigenous people's contemporary response to the "paralysis" produced by the whites' way of treating and interacting with them. These organizations observe that "*They are seeking their self-protagonism in everything related to their own life, and this attitude translates in political, ethnic, identity, cultural and linguistic dynamics; features that are in fact not separable.*"[90] According to Kalish and Unruh, they are trying to recover their own way of being hidden from the white man's surveillance. Their way of being remains valid in everyday life despite the constant pressure on their communities. This is why they consider the communal everyday life as spaces of resistance where indigenous potentialities survive.[91]

The *Sawhoyamaxa* community won the case against the Paraguayan state in 2006 and the land restitution process already started. Time will show to which extent it will empower them to build their own future. La Patria colony's experience over the past 20 years offers the possibility of analysing how the Angaité responded to changes in access, use and management of natural resources inside and outside their colony.

2.2 La Patria colony

2.2.1 Social and ecological setting

La Patria is located in a sparsely populated region of the humid Eastern zone of the Paraguayan Chaco. Despite the augmented non-indigenous population in the region, La Patria is the largest human settlement within a 70 Kilometre radius. *Colonia Ceibo*, located 35 kilometres to the south-east of La Patria, is the closest non-indigenous town and police station.

La Patria is connected to the *Transchaco*[xviii] highway by a dirt road, created by ranch owners, which constitutes the main access to the colony. This road is used to access various ranches, indigenous communities and small non-indigenous towns in the region. A local road commission, presided by a rancher, maintains and manages this access by closing a barrier when it rains in order to avoid transit. Despite all the public money invested and the fact that road level increased considerably along most of the 80 Kilometre from the *Transchaco* to La Patria, a short rain can turn the region inaccessible.

The area has very limited infrastructure due to the combination of frequent corruption and difficult access. Electricity supply was supposed to be granted a few years ago but the intermediate company in charge of it declared bankruptcy and left the project un-concluded. Cell phone companies installed big antennas to provide phone and internet services but cannot operate without electricity. La Patria population communicates using private cell phone antennas in nearby ranches. Two radio stations, *Pa'i Puku* and "ZP30", dedicate some minutes every day to spread personal and communal messages.[xix] There is a first-aid clinic in one of the villages of La Patria, also used by the non-indigenous workers of nearby ranches. However, the nearest medical assistance is located in *Rio Verde* by the *Transchaco* highway. There are primary schools in almost every village of La Patria but the nearest secondary school is also based in Rio Verde. This town also provides the nearby communities with a gas station, a police station, and an

[xviii] *The Transchaco highway is 781 kilometres long. Starts in Asunción and ends in the Bolivian border. It was built with the assistance from the US Agency for International Development. In 1961, the earth road reached Filadelfia (centre of the Mennonites colonies). Asphalting was completed in mid-1980s. J. M. G. Kleinpenning, 2009*

[xix] *In April 2017, some villages were connected to the ANDE (the national electricity supplier). With the electric supply, cellphone and internet service was accessible. Both the electricity and cellphones imply important changes, for example, in the communication among villages and leaders, and also, in the interaction with external actors.*

Anglican residence used by the Angaité. From here, the *Transchaco* reaches Asunción to the South (325 Km) and the Mennonites colonies to the North.

The region where La Patria colony is located is characterised by vegetation associated with the humid Chaco, however, it also includes certain forest formations more typically associated with the dry Chaco.[92] La Patria colony's vegetation cover is characterised by a *quebracho* forest that occupies the relatively high grounds, palm savannahs dominating the slightly lower grounds, grass plains susceptible to flooding covering the lowest areas and gallery forest along the margins of streams[93]. This vegetation sequence is illustrated in figure 4.

Figure 4. Vegetation sequence of the humid Chaco. (Leake, 1998)

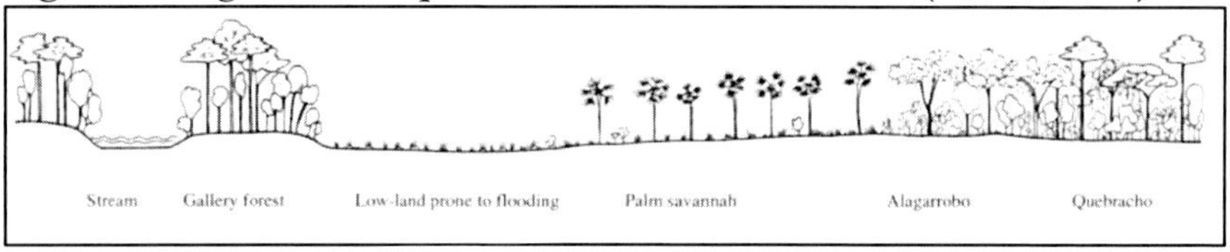

Leake described these vegetal formations as follows. The *quebracho* forest (*Schinopsis balansae* and *Astronium)* form islands of woods within the broad mosaic of palms and grass-covered savannahs. The *algarrobo* forests (*Prosopis)* colonise soils which are subject to extremes of flooding and drought. The low lands, locally known as *bañados* are at the margins of rivers and marshes and may be covered by grass or woody plants. The gallery forest is locally known as *monte alto.* It reaches heights of 20 to 30 meters and is considered as penetrations of the Atlantic tropical rain forest that once covered eastern Paraguay. Finally, the palm savannah (*Copernicia alba)* is the dominant type of vegetation of the humid Chaco. It occupies low lying lands susceptible to floods during 3 to 6 months of the year. Since the introduction of the cattle, the *viñal*, a thick thorn scrub (*Propopis ruscifoglia)* has widespread in the Chaco.[94]

2.2.2 History and current situation

As was already mentioned, La Patria colony was created as part of "*La Herencia*" Anglican project. The 22,520 hectares land property with a mosaic of forest and palm savannahs crossed by streams was purchased in 1983. More precisely, La Patria was established on what once were three paddocks of the 620,825 hectares cattle ranch owned by the IPC.[95] The new colony was represented by a Directive Board (*Junta Directiva*) integrated by 20 leaders. In 1989, nineteen of these leaders, among others, were officially recognised as "indigenous leaders" by the Paraguayan state.[xx]

Leake suggests that the land was selected by the Indian Directive Board of *La Herencia* projects.[xxi] The reasons mentioned were: three of the family groups that moved to La Patria were living close by at the time; the land was located on the border of Angaité and Sanapaná territory; and because land was suitable for agriculture.[96] In August 2015, while conducting my fieldwork, the Angaité celebrated the anniversary of the colony. In his public speech, the leader remembered that in 1983 *"the missionaries bought and returned to the Angaité something that belonged to the Angaité, as stated in the National Constitution and the 904 law".*[xxii]

According to Leake, this colony joined a dozen communities (family groups) "*previously spread over a broad swathe of the Central Chaco.*"[97] Following the settlement pattern suggested by the project, the different groups settled in specific areas of the property chosen by their leaders. As table 4 shows, the re-settled population was living in at least 12 ranches prior to re-settlement. Most of these ranches were located in the land that belonged to the IPC. Some members of these groups were born on these ranches while others previously lived and work elsewhere.

This re-settlement initiated the process of converting "the foragers" into farmers, "*In its heyday, the colony buzzed with the sound of tractors and chainsaws, as agricultural, medical and educational technicians busily sought to get the Indians onto the road to development.*"[98] Re-settled families received food from "*La Herencia*" project for at least six months as part of the "food-for-work" agreement. In other words, La Patria colony was the product of a development project based on what the Anglicans and their advisors judged as the only option for indigenous people.

[xx] *These were: Alejandro Recalde, Lalicio Rojas, Sáncez Cerro, José Landi, Luciano Riquelme, Toti Riquelme, Lino Rolón, Reinaldo Centurión, Julián Rolón, Agüero Fernández, Tomás Fernández, Rubén Fernández, Alón Cabañas, Basilio Sosa, Calí Sanabria, Capata'í Segura, Isasio Segura, Damasio Flores y Alonso Pinasco. (Those underlined are still currently leaders of their villages)*

[xxi] *Local population suggested, however, that the original plan was to purchase estancia Riacho Gonzalez, by far the best located and less affected by floods in the area.*

[xxii] *"Los Misioneros compraron la tierra y le devolvieron a los Angaité lo que era luego suyo, como lo dice la Constitución Nacional y la ley 904." Mr. Santiago, August 2015*

Table 4. Origin of re-settled population[xxiii]

Leader/Leaders	Origin/Ranch	Number of people	Destination
Calí Sanabria, Rogelio Doldán	Riacho González	76	La Paciencia
Marciano Paraguarí, Ayala Chamorro	Buena Vista	131	Carpincho
Damasio Flores, Alonso Pinasco, Pedro Alarcón	Carmelitas	190	Karova Guasú
Juan Carlos Cabañas	Santa Juanita	160	Las Flores
Andres Tomen		No data	La Leona
No data	No data	104	Felix-kue
Zenón Nuñez, Mario Recalde	*Makxawaya* mission	70	Laguna Hū
Luciano Riquelme, Sánchez Cerro	Lapacho	98	Colonia 24
Tomás Fernández, Agüero Fernández, Rubén Fernández	Lucero	32	San Fernández
No data	Isla'í mission	91	Laguna Hu
No data	No data	91	Felix-Kue
No data	No data	32	San Fernández
Julián Rolón, Hilario González	Karanda	190?	Tatarê'i

[xxiii] *The data used in this table was collected during fieldwork and complemented with information provided by Andrew Leake.*

Ten years later, an anthropological research conducted by Leake in La Patria (from 1994 to 1996) showed that households continued to engage in hunting, gathering, fishing and collecting honey. This occurred despite the fact that they had the possibility of developing market-oriented economic activities, which they combined with their wage labour in cattle ranches and horticulture.[99] Leake´s research showed that hunting, fishing and gathering remained an important source of food after resettlement: between 52% and 73% of the protein consumed by households was still derived from hunting and fishing.[100] Considering that these livelihood activities require vast amounts of land, the amount of land used for their livelihood was vastly different to the amount they actually owned.[101] In short, Leake showed that despite colonization traditional livelihood activities persisted. The traditional land-use patterns were significantly altered in this change process. Yet alternative forms of mobility (adoption of horses/donkeys, and the use of camps) emerged and enabled the Angaité to continue hunting and gathering over extensive regions of land.[102]

When Leake conducted his research, the area was relatively unaffected by extensive deforestation. He was right in stating that *"the study area shows that it is prone to the expansion of development pressures in the form of cattle ranching and farming."*[103] Since 1995, a considerable part of the area used by the community to hunt and gather food has been deforested to expand or create new cattle ranches. As the community owned only 10% of the area they used in 1995, deforestation and development of the cattle ranches implied destruction of or denied access to resources which the community depended on in the past and could continue to depend on today. Besides this, regional climate variability challenges their livelihood strategies. The Chaco region experienced, in recent years, significant recurrent extreme events (floods, droughts, storms and wildfires) which resulted in severe environmental, social and economic loss and damages at national and local scale.[104]

2.2.3 Demography from 1995 to 2015

At the time of Leake's study (1994-1996), La Patria consisted of 14 settlements. 10 of them were considered villages and 4 were small hamlets of maximum 3 households. According to a census conducted by colony leaders in 1994, there were 1005 people registered as residents of La Patria which is less than half the total number of people that was settled in the colony in

1983. Leake attributes this change both to the migration of some people to other colonies and to the different methods used to conduct the censuses.

Table 5. Population per village in 1994, 2002 and 2012

Village	Total 1994	Total	Total	Variation %
Laguna Teja	X	50	78	56
Primero de	X	X	X	-
Seis de Marzo	X	X	59	-
Carpincho	113	116	100	(-)13.8
Colonia 24	76	78	144	84.6
Karova Guasú	101	98	66	(-)32.7
Karoa'í	X	121	124	2.4
La Leona	173	233	268	15.2
La Paciencia	128	189	217	14.8
Laguna Hũ	94	156	177	13.4
Las Flores	118	130	218	67.6
Monte Kué	36	44	90	104.5
Paraíso	X	-	127	-
Puente Kaigue	X	43	35	(-)18.7
San Fernández	135	214	230	7.4
Tatarẽ	31	28	31	10
Tres	X	X	18	-
Urunde'y	X	X	47	-
TOTAL	1005	1500	2029	35.2

X=villages that did not exist at the time a census was conducted. (DGEEC 2005; DGEEC 2014)

During his field research, he noticed that the number of people present within the colony varied notably along the year and that only half of the total number of residents was actually present. This is still so, the population increases on weekends, when people working in nearby ranches visit. It is also notable during Easter's holidays which take place in December and January, individuals and families that permanently live on ranches return to La Patria to visit relatives and do maintenance work on their houses.

Since 1996, two National Indigenous Population censuses have been conducted, one in 2002 and another in 2012.[xxiv] According to the 2012 results for La Patria, there were 2029 people, 1020 men and 1009 women. The colony population is currently split into 18 villages, the smallest has 18 people and the biggest has 268 people; on average there are 119 people per village. The population fluctuation can be analysed comparing the three available censuses (see Table 5). From 1994 to 2002 four new villages were founded. Three more were founded from 2002 to 2012. In the same period *Monte Kué*, *Colonia 24* and *Las Flores* villages experienced a population growth of 104,5%, 84,6% and 67,6% respectively. The latter includes new-borns and people moving from another village or community. It is important to consider that in the same period *Karova Guasú*, *Puente Kaigue* and *Carpincho* Villages experienced a population decrease of 32.7%, 18.7% and 13.8% respectively.

2.2.4 Layout of the colony

In 1995, Leake stated that the village settlements have changed since the colony was first established, as a result of population growth and group fission. The 18 currently existent villages are not uniformly distributed within the colony (see Figure 5). Seven are located right on the main dirt road that connects the colony with the *Transchaco* highway. This main dirt road cuts the colony from the Northwest to the Southeast. Entering the colony, visitors traverse *Las Flores* (LF), *La Paciencia* (La Pac), *Urunde'y* (U), *Paraíso* (P), *Puente Kaigue* (PK), *San Fernández* (SF) and *Tres Quebrachos* (3Q). Five of these villages are located on riverbanks or very close to rivers. *Primero de Mayo* (1M), a recently formed village by a stream, is located between *Paraíso* and *Las Flores*. A second series of villages are distributed along the Riacho González, the salty river that crosses the colony from Northwest to East: *Seis de Marzo* (6M), *Karova Guasú* (KG), Karoa'í (Ki), *Laguna Hũ* (LH), *Carpincho* (C), and *Monte Kué* (MK). A third and last series of villages are located along the *bañado*, a water stream connected to the Riacho Paraguay. *La Leona* (L), *Colonia 24* (C24), *Tatarẽ* (T) and *Laguna Teja* (LT) are all located on riverbanks along this *bañado*. The villages located to the East are the first ones to have accessibility problems during the flood seasons. Despite the existence of a net of internal roads that allow the transit of tractors, trucks, pick-ups and motorbikes

[xxiv] *Considering the population mobility these results, and any subsequent quantitative analysis should be considered only partially valid.*

during some months a year. This is not surprising, a considerable amount of the colony surface floods during the wet season.

The location of riverbanks coincides with the Angaité's former traditional residence pattern. Due to the flat topography of the Bajo Chaco region, high grounds are not abundant in the colony.

Figure 5. La Patria colony and its villages

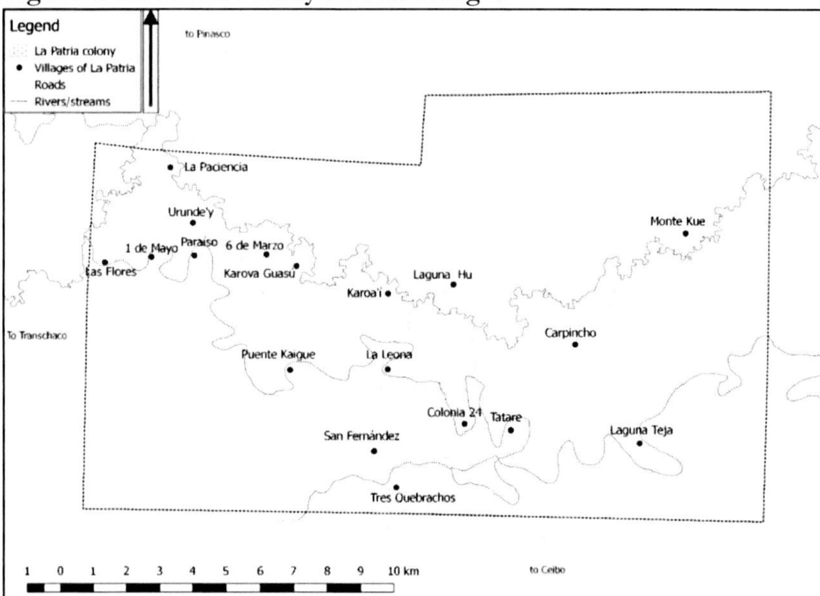

According to Leake, the riverbanks are formed of sedimentary alluvial deposits that not only make the soil slightly higher than the surrounding areas but also more fertile.[105] Riverbank vegetation normally combines gallery forests and small areas of savannah grassland locally known as *espartillares.*[106] The current distribution leaves two main big extensions of the colony with low population density, the extreme Southwest and the North regions.

Box 1. History of existing villages

Based on the memory of some leaders, I reconstructed the history of the current existing villages. Initially, around 1983, 11 settlements were created (*San Fernández*, *Las Flores*, *La Paciencia*, *Laguna Hū*, *Colonia 24*, *Karova Guasú*, *Tatarē*, *La Leona*, *Carpincho*, *Tatarē'i* and Felix-cue). After a short time, the families living in Felix-cue village decided to move to *Sombrero Piri* colony. Eventually, the families living in *Tatarē'i* also abandoned their settlement, some of them moved to *Laguna Hū* village and others left the colony. Around the same time, a group of families coming from *estancia* General Britez founded *Monte Kué*. The number of villages remained stable for a period of about one decade. Around 1995, some families living in *San Fernández* moved and founded *Laguna Teja* and *Puente Kaigue*. Then some families from *Karova Guasú* moved and founded *Karoa'í*. Before 2002, a group of families from *Karoa'í* founded *Paraíso*. Between 2002 and 2005 a group of families from *Laguna Hū* decided to move to *Colonia 24* and founded a sub-village called "25". In 2007, families from *Karova Guasú* founded *Seis de Marzo*. Shortly after that, families from *La Paciencia* founded *Urunde'y* and families from *San Fernández* founded *Tres Quebrachos*. Finally, in 2015, families that originally lived in *Carpincho* and then moved to *Paraíso* decided to found *Primero de Mayo*.[xxv]

2.2.5 Settlement patterns and household composition

In this section I consider the spatial distribution of people within villages. As Kidd points out, anthropologists have often used kinship and marriage to try to find some order in indigenous settlement patterns. Based on this case, I tried to understand the role of kinship and marriage on patterns of residence in La Patria colony.

The 25 households of *Karoa'í* have 22 parent-child links (see Figure 6). Young couples tend to reside close to the wife's parents. This creates three apparently divided but interconnected groups in *Karoa'í*. The central nodes are house 4, house 21 and house 19. The maximal distance between the further distanced located houses of each of these groups is less than 250 meters.

[xxv] *As mentioned by Villagra (2010), settlements are always named by people themselves, like 6 de Marzo, in honour of the leader's birthday. Abandoned settlement are known in the present only if remembered by former inhabitants because their physical remains are erased by the Chaco environment.*

Figure 6. Parent-child relations and sibling's relations in *Karoa'í* village

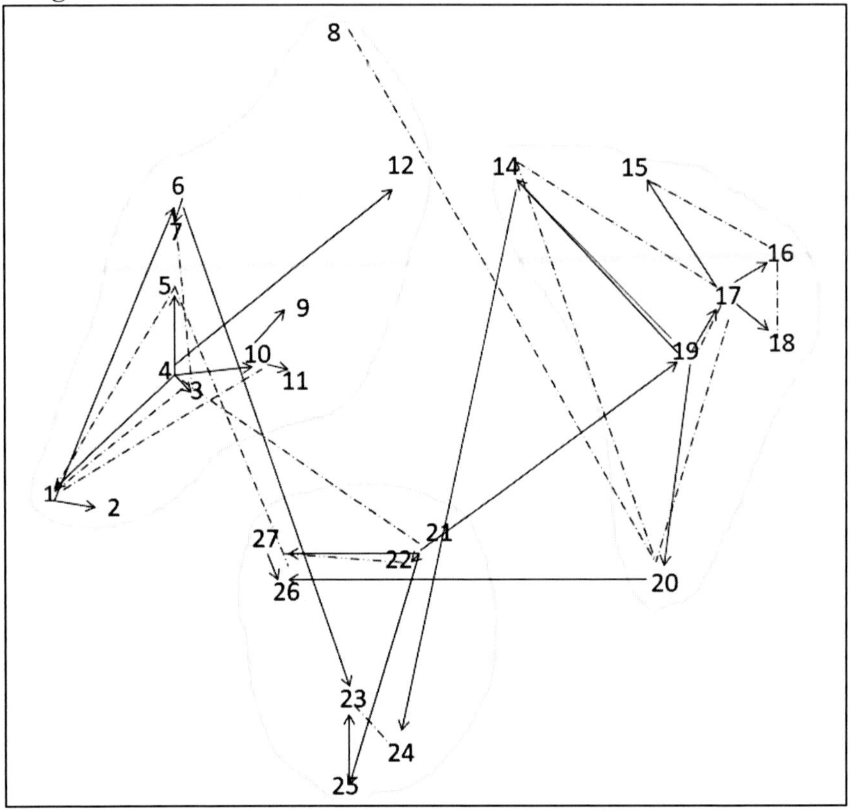

Figure 6 shows with a black arrow the parent-child relations among houses, and with a dashed line, the sibling's relations among houses.

It is important to bear in mind that residential patterns vary through time. For example, the leader's house and his household composition changed since I first visited the colony in 2009. For this reason, it is important to consider the development cycle of the domestic group. This approach allows a more diachronic perception of the residence patterns.

Kidd conducted a similar analysis based on his fieldwork in El Estribo colony, an Enxet colony that shares a close history with La Patria and that belong to the same language family. In La Patria, recently married couples tend to move with the parent of one of the spouses, with a tendency to reside uxorilocally. This allows the recently married daughter to help the mother but also to pool food. When the children grow older, the couple tends to seek more autonomy and/or to look for bigger agricultural plots. This is exemplified by one of the three groups mentioned previously. This group is formed by houses 14, 15, 16, 17, 18, 19 and 20 which are located close to each other in the eastern region of *Karoa'í* village (See Box 2).

Though the study of settlement patterns provides crucial background information with which to understand social relations we also need to understand that settlement patterns and socials relations are not synonymous.[107]

External changes have local impacts. For example, nowadays the presence of unemployed relatives of the indigenous workers at the *estancias* is forbidden.[108] This change translates both into the augmented mobility of indigenous workers, they travel to visit their families at the colonies, and into the employment of young indigenous couples that live permanently at the *estancias*. It is worth noting that a Governmental housing project started the construction of brick houses in some villages of La Patria. Few bad quality houses were completed, hundreds were left unfinished. This was due to corruption, customary in Paraguay for this kind of projects, combined with difficult access to the colony and a particularly strong flood season. However, some are in use and will change settlement patterns because they cannot be moved from one site to another and because people gain social prestige locally by living in such houses.

In order to analyse the household composition and settlement patterns, I conducted a population census, a participatory mapping exercise and a series of interviews. I compared the composition of 19 selected households in 4 villages. My results were contrasted and related with Kidd's findings in El Estribo.[109] According to the Indigenous Population Census 2012 the total population of La Patria colony is divided into 385 households, with an average size of 5,2 members each. I use the household concepts as synonymous of the inhabitants of a physical house, although it could sometimes correspond to the group of people who share the same hearth.

In relation to the household types, I found that out of these 19 considered households only 5 (26,3%) correspond to nuclear families, the rest corresponds to multiple families' households. This is similar to what Kidd found in El Estribo.[110] This contradicts, at the same time, to Renshaw's suggestion that the nuclear family is the most common form of household among the indigenous people of the Chaco region.[111] As Kidd has noted for the Enxet, apparently no formal rule exist among the Angaité regarding post marital residence.[112] According to Renshaw, this is a characteristic of the whole Chaco region.[113]

Box 2. Strong related households in one of the three main groups of *Karoa'í*

Ego, the grandmother (aged 80) is living in house 19 with her youngest son, who is married and has 5 children aged from 8 to 17. Right next door to the south, in house 20, ego's daughter (aged 51) and her spouse live with one 12 years old son. Two other sons already married and left the house (aged 29 and 36). Right to the East to ego's place, in house 17, another ego's daughter (aged 64) and her spouse (aged 61) live with 2 unmarried sons (aged 13 and 16) and one recently married daughter (aged 37) and her spouse. In an independent house, number 18, lives ego's granddaughter daughter, recently married, with her spouse (aged 19). Also in an independent house, number 16, another recently married ego's granddaughter lives with her spouse and one child. A few meters to the Northeast, in house 15 lives another granddaughter (33) of ego, who is married to a man from a different village (aged 40) and has 5 children aged from 6 to 15. Finally, in house 14 lives ego's three times divorced son. He lives with his 22 years old son and her 26 years old married daughter.

While conducting my field research I lived in house 17 for several months. One of the daughters stays alone during the week because her husband works in a nearby ranch and comes back on weekends. Despite the fact that she has her own house, with a kitchen, she ate nearly permanently in her parents' house. A similar situation takes place with another daughter, living in house 18, who is working and living in a faraway *estancia* and comes every 3 or 4 months. When she and her husband came to the village, they stayed in their own house, but ate and shared food in her parents' house. The older daughter who lives in house 15 is different. She and her family are permanently in the village and only eat at her parents' house on special occasions or celebrations. However, material belongings as the ridding seats (*aperos*) and working tools are shared constantly among houses 16, 17, 18 and regularly with houses 15 and 19. Despite the close kin relation and the physical proximity of the houses, a strong social relation with house 15 and house 20 is lacking.

In this chapter, I have shown how kinship and marriage are significant but not determinant factors for the spatial organization of Angaité villages on La Patria. It is important to state here that although household members can co-operate in some livelihood practices (like gardening) they frequently get involved with people from other households in other livelihood practices (like hunting, gathering, fishing). As Kidd pointed out, the groups of people co-operating in some economic activities reflect the personal moods and the preferences of the individuals involved, more than the kin or co-residence among them.[114]

2.3 Conclusion

The Gran Chaco is the second largest biome in South America and encompasses portions of Paraguay, Bolivia and Argentina. The importance of this region, under the increasing stress of demographic expansion and economic exploitation, is lost in the shadows of the Andes and the Amazon. Twenty indigenous ethnic groups classified into six main linguistic families live in the Gran Chaco inhabit this area. Sociologically, these ethnic groups are bounded by a hunting-gathering tradition, common socio-political organizations and socio cosmological aspects. The conquest of the Gran Chaco started in the XIX century and implied the use of terror and violence towards the indigenous population, territorial dispossession, slavery and conversion to Christianism. The different indigenous groups constantly adapted to these circumstances in order to perpetuate their social specificity.

The Angaité are one of the six indigenous peoples that integrate the *enlhet-enenlhet* language family. In 2012, 5992 persons identified themselves as Angaité in Paraguay. They are currently split in seven indigenous communities, out of which, La Patria colony is the biggest one. Only 9,4% of the Angaité speak their Angaité language and the rest speaks "Paraguayan Guaraní".

The Angaité traditional livelihood strategies and land use patterns resemble other native populations of the Paraguayan Chaco. Their livelihood strategy was founded on a diverse and adaptable range of livelihood activities. Their subsistence economy was based on the sharing of food, the free access to natural resources and the autonomy and mobility of the individual as key social practices. The "living well" concept reflects the Angaité philosophy of desirable social practice. Though disrupted by contemporary non-indigenous domination, this ideal is still present.

Colonisation, subsequent privatisation of land and erection of fences was conducted at the end of XIX century. It restricted and eventually halted residential mobility, a key element of the Angaité livelihood strategies. Employment in the tannin industry and cattle ranches offered them a complementary source of subsistence until the IPC crashed when the tannin industry collapsed, in the late 1950s. After that the ranches were subdivided into smaller units.

In 1975, the Paraguayan law established procedures allowing indigenous peoples to make land claims before state institutions. The state, mainly through church-based organisations, tried to integrate the indigenous people into the national society. They encouraged them to resettle in "indigenous communities" and make the transition from "foraging" to farming. Indigenous people did not respond as expected. According to some anthropologist, projects failed to understand the indigenous perceptions and value systems.

In 2014 the indigenous population of the Paraguayan Chaco either owned or had legal rights to occupy a total of 839,313 hectares, not achieving the minimum established by the national law. Besides the ongoing effort to secure more land, indigenous peoples are also facing challenges related to the management of the lands secured in the past decades. Indigenous peoples are seeking to protagonize every aspect of their own lives.

La Patria colony, located in a not densely populated region of the humid eastern zone of the Paraguayan Chaco, was founded in 1983. This colony nucleated dozens of groups of families previously spread in cattle ranches. An anthropological investigation conducted in 1994-1996 showed that despite having the possibility to develop market-oriented economic activities, households continued to engage in hunting, gathering, fishing and collecting honey but combined it with wage labour in cattle ranches and horticulture. Considering that hunting, gathering and fishing requires vast amounts of land, there was a stark difference between the amount of land used for livelihood and the amount owned. The forest cover of the area surrounding La Patria changed dramatically in the past decades. Since 1995, a considerable percentage of the area used by the community to hunt and gather food has been deforested to expand or create new cattle ranches.

According to the 2012 census results for La Patria there were 2029 people, 1020 men, and 1009 women. Currently, the colony population is split into 18 villages, the smallest has 18 people, the biggest has 268 people and on average each has a population of 119. The existing villages are not uniformly distributed within the colony. Current settlement patterns and household composition suggests that kinship and marriage are a significant, but not determinant, factor in orientating the spatial organization of the villages of La Patria.

---

[1] Isabelle Combes, Diego Villar and Kathleen. Lowrey, 2009
[2] Edgardo Krebs and José Braunstein, 2001
[3] Florencia C. Tola, Celeste Medrano and Lorena Cardin, 2013
[4] Isabelle Combes, Diego Villar and Kathleen. Lowrey, 2009
[5] Florencia C. Tola, Celeste Medrano and Lorena Cardin, 2013
[6] Edgardo Krebs and José Braunstein, 2001
[7] Isabelle Combes, Diego Villar and Kathleen. Lowrey, 2009
[8] Florencia C. Tola, Celeste Medrano and Lorena Cardin, 2013
[9] Richard 2011 in Florencia C. Tola, Celeste Medrano and Lorena Cardin, 2013
[10] Florencia C. Tola, Celeste Medrano and Lorena Cardin, 2013
[11] Florencia C. Tola, Celeste Medrano and Lorena Cardin, 2013
[12] Ronaldo Unruh, 2015
[13] Florencia C. Tola, Celeste Medrano and Lorena Cardin, 2013
[14] Alain Fabre, 2016
[15] Rodrigo Villagra, 2010
[16] Unruh and Kalish 2003 in Alain Fabre, 2016
[17] Rodrigo Villagra, 2010
[18] DGEEC, 2014
[19] Alain Fabre, 2016
[20] Alfred Métraux, 1946
[21] Rodrigo Villagra, 2010
[22] Rodrigo Villagra, 2010
[23] Alfred Métraux, 1946
[24] John Renshaw, 2002
[25] Warbroocke Grubb, 1911: 189
[26] Rodrigo Villagra, 2010
[27] Ibid.
[28] Rodrigo Villagra, 2010: 109
[29] Andrew Leake, 2006
[30] Rodrigo Villagra, 2010
[31] John Renshaw, 2002; Andrew Leake, 1998
[32] Andrew Leake, 1998
[33] Andrew Leake, 1998; Volker Von Bremen, 1987
[34] Susnik 1981 in Andrew Leake, 1998
[35] Braunstein 2005 in Rodrigo Villagra, 2010
[36] Rodrigo Villagra, 2010: 101
[37] Andrew Leake, 2006
[38] Andrew Leake, 1998; Alfred Métraux, 1946; Rodrigo Villagra, 2010

[39] Rodrigo Villagra, 2010
[40] Ibid.
[41] Ibid.
[42] J. M. G. Kleinpenning, 2009; Carlos Pastore, 2008; Rodrigo Villagra, 2010
[43] Rodrigo Villagra, 2010
[44] J. M. G. Kleinpenning, 2009
[45] Andrew Leake, 1998
[46] Rodrigo Villagra, 2010
[47] Ibid.
[48] Valentina Bonifacio and Rodrigo Villagra, 2016; Rodrigo Villagra, 2010
[49] Rodrigo Villagra, 2010
[50] Rodrigo Villagra, 2010
[51] Rodrigo Villagra, 2010
[52] Córdoba, Lorena; Bossert, Federico and Richard, Nicolás, 2015
[53] Grünberg et al., 1978; Fogel, 1988; and Chase Sardi et al., 1990 in Andrew Leake, 1998
[54] Rodrigo Villagra, 2010
[55] Ibid.
[56] Valentina Bonifacio, 2013b
[57] Andrew Leake, 1998
[58] Stephen W. Kidd, 1992
[59] Andrew Leake, 2006
[60] Andrew Leake, 1998
[61] Rodrigo Villagra, 2010
[62] Stephen W. Kidd, 1995; Rodrigo Villagra, 2010
[63] Andrew Leake, 1998: 78
[64] Belaieff, 1946 in Andrew Leake, 1998: 78
[65] Andrew Leake, 1998
[66] Susnik 1977 in Rodrigo Villagra, 2010
[67] Andrew Leake, 1998:80
[68] Grünberg et al. 1978 Andrew Leake, 1998: 80
[69] Wallis 1986 in Andrew Leake, 1998
[70] Andrew Leake, 1998: 81
[71] Stephen W. Kidd, 2000b: 1
[72] Stephen W. Kidd, 2000a
[73] Andrew Leake, 1998: 89
[74] John Renshaw, 2002: 7
[75] Ibid.
[76] Volker Von Bremen, 1987
[77] Ibid.

[78] Hannes Kalish, 2000
[79] Mario Blaser, 2004a
[80] Stahl 1993 in Stephen W. Kidd, 2000b
[81] Oscar Ayala, 2016
[82] Rodrigo Villagra, 2014
[83] DGEEC, 2015
[84] Andrew Leake, 2006: 1
[85] Ibid., 27
[86] Andrew Leake, 2006
[87] Ibid. , 27
[88] Alexandre Surrallés and Pedro García Hierro, 2005
[89] Ibid.
[90] Ronaldo Unruh, 2015: 1
[91] Ibid.
[92] Andrew Leake, 1998: 35
[93] Ibid.
[94] Ibid., 35-39
[95] Ibid.
[96] Ibid.
[97] Ibid.
[98] Ibid.
[99] Ibid.
[100] Ibid.
[101] Ibid.
[102] Andrew Leake, 2006
[103] Andrew Leake, 1998: 141
[104] UNDP, 2012
[105] Andrew Leake, 2006
[106] Rodrigo Villagra, 2010
[107] Stephen W. Kidd, 2000b
[108] Rodrigo Villagra, 2010
[109] Stephen W. Kidd, 2000b
[110] Ibid.
[111] John Renshaw, 2002
[112] Stephen W. Kidd, 2000b
[113] John Renshaw, 2002
[114] Stephen W. Kidd, 2000b

3 Cosmographies, territories and land use

3.1 Introduction

In this chapter, I explore how the Paraguayan state *de facto* allowed and promoted resource use practices in the cattle ranches, over the last two decades, which were incompatible with certain Angaité livelihood practices. This allowance contradicted and defied the state's *de jure* stand regarding Angaité territorial use rights over "La Patria territory". The Angaité of La Patria responded to this situation adapting their land use patterns. I consider that land use and livelihood practices are central cosmographic practices. They contribute to the production of a particular place or territory.

In the first section, I contrast the Angaité cosmography with the "development" cosmography of the state and the cattle ranchers. I explore how the Angaité relate to their biophysical environment through ritual practices, like the *choqueo* and shamanism, as well as through livelihood practices, like the *marisca* (hunting, fishing and gathering). These practices include non-human and "not so human" beings (like the Paraguayans)

In the second section, I focus on recent territorial dispossessions and land use pattern changes. First, I briefly outline some methodological considerations that allowed me to contrast the land use in 1995, as described by Leake, with the situation I witnessed in 2015. Following that, I describe how the state proceeded to issue environmental permits. Authorizing resource use practices made it difficult or impossible for the Angaité to continue implementing livelihood and land use practices. Then, I present the results of the participatory mapping of the land use patterns in 2015.

In the last section, I present the Angaité's opinions on recent land use and territorial relation changes. I do so in order to prove that the current territorial struggle is more than just a struggle over resources.

But first let me stress one aspect that I consider crucial in the events considered in this chapter and this dissertation. I already mentioned some of the international and national regulations regarding the rights of indigenous peoples in Paraguay. In particular, the article 64 of the Paraguayan national constitution says: *"The indigenous peoples have the right to communal ownership of the land, in an extension and quality sufficient for the preservation and the development of their particular ways of life."*[1] In the last pages of his PhD dissertation, Leake quotes Villagra when he states that "*Political inertia and economic greed make it unlikely that the Indians in the Chaco will ever gain legal rights of ownership over territories which are sufficiently large enough to encompass all the natural resources they require to preserve and develop their respective lifestyles.*"[2] Reflecting on this, Leake concludes that the indigenous peoples' rights under the Paraguayan law were "*little more than a false promise*" and that an alternative approach to the current model was required.[3] The strategy used in the last decades by indigenist institutions to secure land for indigenous communities shows that this *status quo*, the lack of recognition by the state of the territories currently or historically used by indigenous peoples, has been somehow accepted. In other words, indigenist institutions focused on securing small portions of land, postponing or relegating the effective recognition of their use rights over more extensive territories.

An example of an alternative approach can be found on the legal argument made by TierraViva in an Action of Unconstitutionality presented to the National Supreme Court in 2008.[4] The interpretation of article 64 of the Paraguayan national constitution focuses on the principles of the international treaties on indigenous peoples' rights ratified by Paraguay. According to this, the extension and quality of land should be defined based on the declared objective of article 64: the preservation and the development of the indigenous particular ways of life. Following this, it is obvious that in the case of the indigenous people of the Chaco this "particular ways of life" includes their traditional livelihood strategies and land use patterns. The second part of article 64 mentions: "*The removal or transfer of the indigenous people from <u>their habitat</u> without their express consent is prohibited.*"[5] The term "habitat" refers to a place that has the appropriate characteristics for a person or a group to live. In this case, the "traditional habitat" should be defined according to the international legislation on the subject. This is due to the fact that the extension of land a community uses or has been using to conduct the livelihood activities contributes to the preservation and development of their particular way of life.[i] This definition of the "traditional habitat"

[i] *Article 13 of the ILO 169 Convention says: "In applying the provisions of this Part of the Convention governments shall respect the special importance for the cultures and spiritual values of the peoples concerned of their relationship with the lands or territories, or both as applicable, <u>which they occupy or otherwise use</u>, and in particular the collective aspects of this relationship. The use <u>of the term lands in Articles 15 and 16 shall include the concept of territories</u>, which covers the total environment of the areas which the peoples concerned occupy or otherwise use."* ILO, 2003

includes the rights to use resources located on land properties not owned by them.[ii]

As we know, most villagers of La Patria lived in or close to the colony before it was created. By this I mean that they continued to live in and use what according to the law is their "traditional habitat". Other villagers came from areas located further away from La Patria, but even in those cases, all of them were living in what is considered to be the traditional Angaité territory. As I am focusing on the resources and lands used by the Angaité of La Patria in 1995, disregarding their origin and past land use, I refer to these lands and natural resources as "La Patria territory". "La Patria territory" should not be confused with the Angaité traditional territories, as represented in Figure 3.

The villagers of La Patria colony were using around 209.000 hectares of land in 1995. These 209.000 hectares constituted "La Patria territory" in 1995 (see Figure 10). This, as we know, includes lands privately owned by third parties. There is an overlap of rights on those areas. The Angaité have the right to access and use the resources for their livelihood activities. The ranchers, owners of the land, have private property rights and use rights. Thus, "La Patria territory" is *de jure*, a territory that is subjected to laws that establish specific rights and obligations. The fulfilment of those rights, and compliance and enjoyment of these territorial rights are an obligation of the Paraguayan state.

The ILO Convention 169, in article 15, states that the right of indigenous people over the natural resources located in their territories should be safeguarded, including the right to participate in the use, management and conservation of these resources.[6] This implies that *de jure* the management of "La Patria territory" should consider the needs and interests of both parties: ranchers and the Angaité. For example, if a ranch that wants to deforest to create new pastures requests permission from the SEAM (National Environmental Agency), the SEAM should make the appropriate consultation in order to consider the Angaité rights and interests. In this way,

[ii] *Article 14 of the ILO 169 states: "The rights of ownership and possession of the peoples concerned over the lands which they traditionally occupy shall be recognised. In addition, measures shall be taken in appropriate cases to safeguard the right of the peoples concerned to use lands not exclusively occupied by them, but to which they have traditionally had access for their subsistence and traditional activities. Particular attention shall be paid to the situation of nomadic peoples and shifting cultivators in this respect. Governments shall take steps as necessary to identify the lands which the peoples concerned traditionally occupy, and to guarantee effective protection of their rights of ownership and possession." ibid.*

those areas and resources that are crucial to the Angaité could be protected, including those areas in the 25% forest reserve that ranchers are forced to maintain by law. I am not suggesting an alternative approach here; I am describing what *de jure* exists and is supposed to be implemented. As we will see, *de facto* the Angaité territorial right to access and use these resources is not respected but also completely ignored by the national authorities and the powerful local ranchers.

It is important to remember that Angaité and cattle ranchers do not only compete over the use of same resources, they embody contrasting cosmographies and nature regimes. "La Patria territory" has, thus, two dimensions. On one hand, it is a land area, over which the Angaité of La Patria have use rights. On the other hand, it refers to a fabric of social relations between different beings that constitute a territory. In order to understand the social relations involved in the second aforementioned dimension I contrast the particular characteristics of the Angaité and the ranchers' cosmography, in the following section.

3.2 Different cosmographies and "nature regimes"

When I visited La Patria for the first time in April 2009, I was invited by a friend who was working for a Paraguayan NGO called Grupo Sunu. This NGO was supporting, by offering food donations, the realisation of a *choqueo* that was part of the "Indigenous people's day" celebration, on April 19th. Hundreds of families from different villages of La Patria, and even from other communities, gathered to celebrate. Small fires illuminated the families that were camping around the central circle where the musicians (*músicos)* played the drum and sang. People were expressing their happiness in different ways. Mr. Santiago, leader of *Karoa'í* village approached me and said: "*This is our culture*". The *choqueo* ritual had been interrupted for at least two decades and was, not without resistance, reactivated in April 2006 in *Karoa'í* village. To celebrate the reduced influence of the Anglican missionaries coupled with the increased involvement of leaders and elders in a cultural and linguistic recovery project, some leaders decided to organise a celebration. This celebration basically consisted of a *choqueo* and a traditional food celebration. A video documentary made by Grupo Sunu in that occasion contains interesting expressions that hint at the Angaité cosmography and the situation of La Patria in 2006.

In the documentary, an elderly man speaks about the celebration:

"I want to speak about the dances that we used to perform. I got to see them. The drums are beating. The little bells also. There is a lot of noise. The celebration is prepared when a young girl turns fifteen. The women set their beds on the floor and dance till dawn, until sunshine. We used to live like that, there was food in abundance. Of every kind!..."[7]

In the same context, a woman speaks about the times when the Angaité used to live in the *estancias*. She also mentions the abundance of food:

"When we lived in (estancia) "14 de mayo" there was abundant food. We didn't lack anything. (…) Now when the weather is bad, we don't have anything to feed our children. And it really hurts to see our children and families without anything to eat. And their parents look for food inside the colony's property. But our area is not fertile any more. We need to change. And we ask in our own language for your support with food. (…) This is how we used to live, now we just survive. Our grandchildren were not fed with this real food."[8]

A few minutes later, while presenting wild plants that the Angaité ate until a few decades ago and that remain unknown to the younger generations in 2006, Ricardo Ramirez implicitly revealed how food and non-humans were connected. He said:

"We used to eat this fruit of the 'bromelia'. But now it does not bear[iii] *as it used to. Our Father*[iv] *is eliminating it fruits, and the same is happening to us. We have to find new ways to subsist now. (…) It is also hard to find 'kathan' in the colony's property. It is not growing well anymore. (…) These were our real food items and we are not eating them anymore. And where did they come from? These foods were given to our ancestors by Our Father. He spoke in our language and said that we should eat it. But now it seems that Our Father is leaving us without food. (…) These foods are finishing and they will end up disappearing."*[9]

[iii] *"Ya no rinde"*

[iv] *Our Father is a new auxiliary spirit incorporated in the process of constant transformation of shamanism since the colonization process began. See Villagra, 2010: 199*

What these three statements have in common is that they refer to a time of abundance (related to the will of a non-human agent, in this case it is Our Father) contrasted with the scarcity of present time. These statements evidence that the Angaité, like many other indigenous groups, do not make a clear nature-culture divide or split their existence into religious, political and economic domains.

3.2.1 Understanding the Angaité cosmography

For his land use study among the Angaité, Leake used a territoriality definition suggested by Casimir in 1992. This definition states that *"Human territorial behaviour is a cognitive and behaviourally flexible system which aims at optimising the individual's and hence often the group's access to temporarily or permanently localised resources, which satisfy either basic and universal or culture-specific needs and wants, or both, while simultaneously minimising the probability of conflicts over them."*[10] This definition, while adequate for Leake's research objectives, is based on the Western naturalistic ontology. It omits crucial aspects of the Angaité territoriality that are based on a different nature regime. Escobar mentions that a defining feature of the "organic nature regimes" is that nature and society are nor separated ontologically. They are often based on the links between the biophysical, human and supernatural domains. This continuity, he adds, is established through rituals and practices embedded in social relations. Territory, in this case, is understood as a multidimensional entity that results from many types of practices and relations.[11]

This resonates with Von Bremen's argument about the inexistence of a conceptual division of culture and nature among the indigenous peoples of the Chaco. According to him, among these peoples, there is a shared belief that resources for survival are abundant. He argued that despite their profound knowledge about nature, based on their observation and practical experiences, they needed particular techniques to appropriate these resources. For all these chaquean indigenous groups, these magic (rituals) and practical techniques (like those involved in hunting) have a mythological fundament, "*all natural and cultural phenomenon have a spiritual aspect, they possess their own individual personality… and even their own will.*"[12] Through the magic techniques, "*a contact with the spirit of different important being is established, in order to appropriate the food.*"[13] Von Bremen adds at the end of his description that in order to successfully appropriate the existent resources it is essential to know their characteristic qualities and to establish contact with the spirits of each phenomenon in order to make the most of this relation.[14]

For the Angaité, people, animals, plants, and other non-humans form part of a wider collective. The ethnography of indigenous peoples shows that other societies have other ways of conceiving the "social collective" that unites humans and non-humans.[15] Bruno Latour conceives humans and non-humans as actants, this term includes spirits, animals, ideas, institutions, politics, laws, technologies, techniques, materials, etc.[16] In a similar line, Philippe Descola argued that animists attribute animals (and other non-humans) with intentionality, self-consciousness, emotions and other characteristics identical to those of human beings.[17] I suggest that for the Angaité, as did other scholars regarding other chaquean and Amerindian societies, what Western science calls nature constitutes part of a complex network of social interactions in which man is no more than one actor amongst many others.[18] This is important to understanding the Angaité territoriality. For them territory is far more than an area of land for subsistence and social reproduction limited to a local group that exercises control over the space. In fact, territory belongs to the social sphere.[19] In order to shed some light on the Angaité cosmography, in the next section, I focus on two types of cosmographical practices: the ritual practices (like the shamanism and choqueo) and livelihood practices (like the *marisca*).

3.2.1.1 Shamanism among the Angaité

Valentina Bonifacio and Rodrigo Villagra relate the non-humans to a series of practices that the Sanapaná, Enxet and other groups closely related to the Angaité call "shamanism".[20] The Angaité frequently use the word *apyoholhma* to refer to the shaman.[21] According to Kalish, in its original semantic construction it means "*he who knows the space/forest*" but in its current use it means a "*person with power*".[22] In communities under heavy Christianised influence, people refer to the *apyoholhma* as "healer" or "old man".[23] Among the Angaité and the Maskoy who speak the Paraguayan Guaraní language, is also frequent to call them *mba'e kuaa*, "*someone who knows about things*" or *paje*, "*sorcerer/wizard*".[24]

Shamans in the Chaco fast and ingest rotten plants to learn the ability to travel in the spirits world "*and to get in contact with the different types of spirits that are the owners of different kinds of non-human beings: plants, animals, natural elements and objects.*"[25] According to the same scholar, that based her statements on Pastor Arenas and Stephen Kidd work, shamans, in case of necessity, could "use" the particular abilities of each of these spirits. Shamans, usually but not exclusively men, could, for example, ingest some plants to acquire the ability to act on people's *wa'xok; py'a* (stomach/entrails) to make them more or less generous.[26]

For the *enlhet-enenlhet* (Angaité, Sanapaná, Enxet, Enlhet, Guaná, Maskoy) the cosmos is not only inhabited by human beings, animals and plants, but by many other type of beings, namely, those similar to "persons" and those simply classified as "things".[27] According to Villagra, in the cosmology of these Chaco peoples, the "human mode" (person/people) occupies the referential position as subject. It clearly distinguishing from "non-human" modes (animal, plants, things) and the "not so human" and "meta-human" modes of foreigners and some types of spirits respectively[28]. Different types of beings live in the different levels of this multidimensional cosmos, however, the shamans are the only ones able to travel to other levels to find these *apmovana* (auxiliary spirits).[29] If a shaman succeeds in his travel, he finds these spirits in their own places, they appear in a recognisable shape that can be zoomorphic or anthropomorphic and are willing to help their human partner to master the "things" and "powers" (*poderes*) controlled by them. These "things" can be animals, plants, beings of monstrous appearance, manufactured objects like money or planes, etc. The Angaité translate the concept of *apmovana* to the Paraguayan Guaraní *ita'yra (his son), imba'ekuera (his things), che socio* (*my partner*) or with the Spanish word *espíritu* (*spirit*).[30]

A related issue mentioned by Kidd is that the Enxet shaman's capacity to communicate with these auxiliary spirits "*was used to protect the communities and to provide them with food and other material benefits.*"[31] This is why the Angaité relate the rituals with times of abundance. They attribute the shortage that characterises current times to the reduction in serious rituals which could result in a reduction of wild animals and bad harvests.

3.2.1.2 From the Vaingka to the Choqueo

The *choqueo*, widely practiced in many indigenous communities (not only Angaité), is commonly described as a group of people dancing around a singer who plays a water drum. The circle is initially solely composed of males but women carefully chose where to enter for a few minutes before leaving the circle again.[32] The main singer receives some shamanic training that involves fasting in the forest and drinking the water contained in the hollowed palm trunk in order to learn, from the "owner" of the song, how to sing at a high performance level.[33] However, Villagra warns, "*this type of training and subsequent acquired singing capacity did not alone guarantee further capacities associated with more powerful shamans.*"[34]

The *choqueo* is widely known as a "traditional dance" of some indigenous groups of the Chaco region. Today, it is also practiced in the context of Indigenous Organization meetings or demonstrations related to the territorial and land restitution struggle. As Bonifacio and Villagra argue, "*the complex set of relations that link together the people song with the 'earth beings' or auxiliary spirits remains hidden*" to the non-indigenous audience, that usually just consider it a cultural expression.[35] This "traditional dance" and the *vaingka,* the ritual celebration of which it was part of, have been transforming since the colonization of the Chaco. This is precisely what Villagra aimed to understand in his article entitled *From the Vaingka to the Choqueo.* In this article, he evidences that the *choqueo* has replaced the *vaingka* since the Paraguayan colonization of the Angaité territory (1880-1943) and the subsequent development of the cattle ranches. The *vaingka* had a ritual character and involved powerful shamans as singers. The shaman song is an exercise of the powers acquired by the shaman from it auxiliary spirit from the forest, the fauna or the "others".[36] The *vaingka* included a feast for the guests, thus, it was only possible during the rainy season (November to February) when there was an abundance of food. This abundance of food made happiness possible and boosted the "buen vivir" or "sharing" with the guest, one of the central aspirations of the collective Angaité life.[37]

The development of the cattle ranches, Villagra argues, constituted the new socio-economic and geographical pattern of reproduction of the indigenous population. This change and transformation process led to the substitution of the *vaingka,* which once made the interrelation between pre-colonial villages and groups possible, by the *choqueo.* The latter became a way to process the alterity and ethnic asymmetry of the non-indigenous Paraguayan or *Chaqueños* with whom the Angaité coexisted in the cattle ranches.[38] In the ranches, the indigenous populations used to organise *choqueos* during the weekends, and the non-indigenous populations used these opportunities to get in touch with indigenous women. In order to do so they had to respect the ritual protocol of the *choqueo* and compete, at an equal level, with their indigenous co-workers. The *choqueo* forced the Paraguayans to respect the relations the indigenous population defined, which were based on an equity that contrasted with the asymmetrical relations of the everyday work relations.[39] Villagra uses this analysis to explain the changes experienced by the Angaité without denying some continuity and keeping in sight the asymmetric relations in which they are involved. He concludes that the *choqueo,* which fulfils the same purpose as the *vaingka*, is part of the Angaité strategy through which they process the alterity of the non-indigenous with whom the Angaité socialize, work and sometimes dispute.[40] The ways in which these rituals have been changing provide important information about the changing Angaité cosmography.

3.2.1.3 Marisca as a relation with non-human beings

In a conversation that Florencia Tola had with a Qom leader from the Argentinian Chaco, he stated: *"In the Qom culture*[v] *every habitat has its own owner. This is why our ancestors prayed every time they entered the forest. They prayed to return with abundant food. It was by doing so that they received everything."*[41] In the same vein, the Angaité refer to different kinds of places and use to hunt or gather in the same places because they "know" a particular site. If every forest has an owner, how are they affected by the process of deforestation? The Angaité cosmology and shamanism have innovated and likewise incorporated foreign auxiliary spirits and new techniques, argues Villagra.[42] He adds that this allows for the adjustment of the explanatory force of such practices under the current everyday conditions. Some Enxet elders who worked with Branislava Susnik, states Villagra, *"asserted that the owner of the animals have reacted themselves to the new historical and ecological conditions."*[43] These elders, told Susnik that the massive introduction of cattle and the "noise" produced, *"upset the animals and their spiritual 'owners', henceforth the former have gone into the deep forest where they are hidden by their 'owners'."*[44] Tola posed the same question regarding deforestation to the same Qom leader. His response was nearly identical to the one formulated by the Angaité a few decades earlier. Finally, the Qom leader added, *"As the habitat is disappearing, the communication reduces and becomes scarce. (…) …through the activity of marisca we use to communicate with them (the owners), but as the territories are privatised, we have little access to the practice of marisca and to the relation with these beings."*[45] Tola concludes that when the Qom relate to what for us is just a resource, visible and natural, other beings emerge. These beings, she argues, do not exist not in the external space but in the indigenous *praxis.* Therefore, the *marisca*, synthetises the visible and the invisible, the spiritual and the material, the human and non-human.[46]

We can close this section stating that the Angaité cosmography is better understood using a territory definition suggested by Surrallés and García

[v] *All quotes form Tola's work are my own translation.*

Hierro. From their perspective, the territory, *"far from constituting a geometrical area framed by physical landmarks that separate and demarcate it, an indigenous territory is simply the consolidation of a very specific and singular fabric of social ties between the different beings that make up that environment."*[47]

3.2.2 The "development" cosmography of the cattle ranchers

In this section, I characterise the cattle ranchers as a central part of the state/modern cosmography deployed in the Angaité area since the beginning of the twentieth century. This cosmography can be described as what Escobar terms "capitalist nature regime".[48] For Escobar, this regime resulted from a particular phase of history: patriarchal capitalist modernity. A fundamental feature of this regime is the conversion of nature into resources for humans to be used as we wish. He relates the capitalist modernity to the development of a governmentality based on expert scientific knowledge. The capitalist nature regime, like other regimes, is historical. The ways nature or the environment and the social relations were conceived in the cattle ranches have been changing. I will try to show that an important change occurred at the beginning of the 1990s. It made nature, increasingly conceived as uniform, manageable and harvestable. As a result, mentioned by Escobar, it is becoming an obstacle for the capital accumulation itself.

The cattle ranches started to be installed in the Chaco a few decades after the quebracho industry began to function. The International Products Corporation (IPC) had, around 1918, three well established ranches: *Kilometro 80* (later known as *Tuparendá*), *Riacho Gonzalez* and *Laguna Tigre.* When the tannin industry collapsed these and other ranches were subdivided into smaller family-run units or *estancias.* Some of this families argue that they were the first settlers of the Chaco and claim a civilizatory role in "*opening the doors for the economic progress in the Chaco inland.*"[49]

The family-run *estancias* in the surroundings of La Patria practiced, during a couple of decades, manual forest clearance. It contrasted the mechanised forest clearances practices that became common in the mid-1990 and peaked around 2012. During the 1980 and until 1995, the forest clearance pattern in this region consisted of small and isolated patches of less than 1000 hectares.[50] During this period, ranchers used to clear small areas of their properties to provide managed grazing sites mainly used in the winter months. During the rest of the year, the palm savannahs and *bañados* provided sufficient pasture for cattle. These clearances were typically done by hand:

cutting trees above the ground, burning them once dry and leaving the root structure in the ground. Unlike mechanized forest clearance, this manual practice left the soil more protected against erosion.[51]

Leake also mentions some ecological impacts of ranching in the Chaco region. During the 1990 it was already clear that uncontrolled grazing was "*the most destructive influence that has yet affected the arid ecosystems of the Chaco.*"[52] Based on the relatively low impact of the above-mentioned hand clearance practices in the neighbouring cattle ranches of La Patria, Leake argues that degradation rather than deforestation was the main ecological impact in this area. He clearly states: "*although large areas of the Paraguayan Chaco have been cleared, deforestation is not yet a major problem within the study area.*"[53] Keep in mind that some landowners were just real estate speculators and were not conducting any economic activity in the property each legally owned. He also noticed that this situation was rapidly changing at the time of his field study, as "*the division of ranches into smaller units has meant ranchers have increasingly resorted to clearing forests.*"[54]

Besides deforestation and degradation, Leake mentions another environmental impact that would grow over the next years. He mentions that the upper reaches of the drainage basins of the streams which run through La Patria area coincides with the Mennonite settler's colonies that constituted the most extensively deforested area in the central Paraguayan Chaco, at that time. He argued that topsoil erosion increases silt flow.[55] However, this was not the most severe impact on local water supplies. Around 1995, damming streams to create small dams was already a widespread practice. These small dams not only interrupted the water run-off but also increased salinity as the result of the hydrostatic pressure caused by water in the reservoir.[56]

3.3 Recent territorial dispossession and contemporary land use

3.3.1 Land use of the Angaité of La Patria in 1995

In 1995, Leake used the villages as basic social unit of land users due to the communal use of lands among the Angaité. The entry point to his study of land use patterns was the land used for hunting. These areas tended to encompass other areas required for gathering and fishing. Using a participatory methodology, Leake asked members of all villages to draw maps of the hunting areas outside the colony. Most of them drew diagrams of the ranches where such areas were located. This was complemented with interviews. In this way, he acquired a general knowledge of the location of each village's hunting zone. As the Angaité named ranches to design the areas used for hunting, Leake used these ranches to create a map of the overall areas used by all villages.

Figure 7. Overall area containing hunting sites used by the various villages of La Patria in 1995/96.

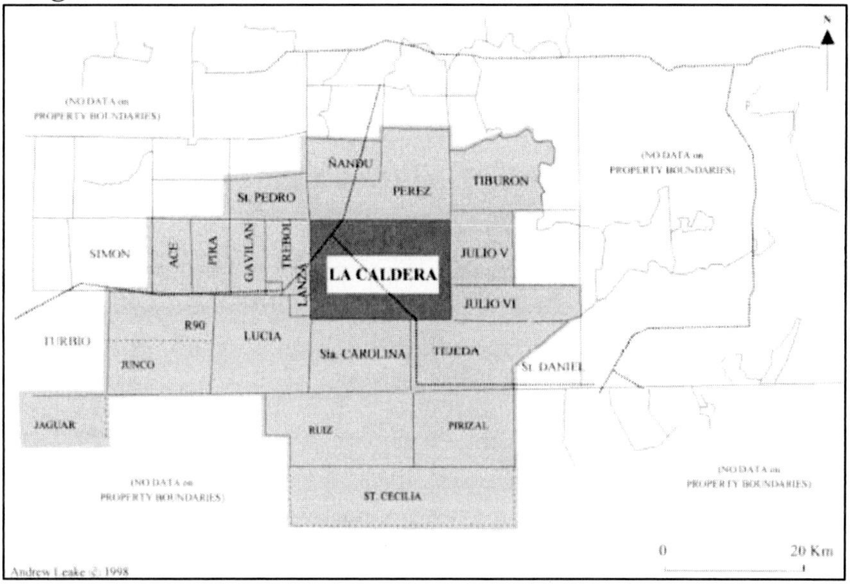

„La Caldera" = „La Patria". Source: Leake, 1998

The map shows the ranches periodically used by the villages of the colony in 1995. To maintain anonymity, Leake changed all ranch names and some property boundary lines. The ranches lie within an area of approximately 60 by 40 kilometres, and cover an area of around 240,000 hectares. This area was ten times larger than the land the Angaité owned.[57] Leake defined the Angaité's land as their 'home range', an "*area within they move on a periodic basis for the purpose of subsistence procurement.*"[58] He found that La Patria colony provided only a fraction of the land the Angaité actually used in their livelihood activities. Thus, there was an extensive overlap between lands used by different groups. This means that "*there [was] a major discrepancy between the amount of lands that the Indians own[ed] and the lands that they actually use[d] for their subsistence.*"[59]

The Angaité only used selected sites within the ranches. Some ranches were no longer used by the Angaité due to the fact that they had been totally deforested. Leake remarks that some of the areas used for hunting had only recently been used by the Angaité. In those areas, there was evidence (honey extraction holes in trees) of use by indigenous people prior to the establishment of La Patria. Leake identifies the following land uses in "core areas" of intensive use: a site of residence; a place for cultivation and an area to procure water and firewood; and a "periphery" of land used on a more extensive basis including sites for the procurement of wild animals, fish, wild fruits, honey and materials for construction.[60] He argues that these areas play a complementary role within the subsistence economy.

> Leake concluded that the land of La Patria colony "*provides local villages with secure access to their respective core areas of land use, but significantly fails to ensure their access to most of the sites used within its peripheral zone, because these lie inside lands owned by cattle ranchers.*"[61] He also states that "*the current levels of wellbeing are intricately linked to people continuing to have equal access to natural resources over a broad area of land.*"[62]

Leake identifies a pattern, villages where some members sought wage labour were shown to be associated with particular ranches. Showing how the access to these areas was slowly becoming more and more complicated he adds: "… *it appeared that in many cases the right to hunt on a ranch was conditioned by whether or not Indians had some form of link to the ranch owner or their care taker.*"[63] Revealing a strategy that would become mainstream, he points out that "*In some instances it was clear that people undertook paid jobs primarily to ensure access to hunting grounds, a fact which may help explain why Indians seek work even for very low pay and frequently atrocious working conditions.*"[64]

This evidences that hunting was still possible in 1995 because La Patria was situated within a broader "*relatively untouched region which still provides a large and non-fragmented habitat for wildlife.*"[65] It is also clear that the Angaité way of life and wellbeing was already threatened by the process of intensification of neighbouring cattle ranches as well as by subsequent deforestation and degradation of the forest that the Angaité depended on.

3.3.2 State natural resource management practices and the recent territorial dispossession

The Angaité are facing a new process of territorial dispossession that started in the mid-1990s and reached its peak around 2012. The soy boom experienced by Paraguay in the last decades indirectly impacted the Angaité area by pushing the cattle ranches into the Chaco region. The expansion of old cattle ranches and the development of new ones explain the incredibly high deforestation rates reached in the Paraguayan Chaco in the past years.

Paraguay land distribution is considered the most unequal worldwide, with a Gini coefficient[vi] of 0,94 in the Chaco region and of 0.89 in the Eastern region.[66] According to the last Agricultural Survey, 2.6% of the agricultural properties use 85.5% of the agricultural land.[67] Nationwide land concentration and creation of new large estates were observed: agricultural properties of 100 to 499 hectares increased 34.9% and properties of 500 or more hectares increased 56% respectively in the past 20 years.[68]

In 2008, 25.3% of the agricultural land was owned by foreigners: 4,792,528 hectares were owned by Brazilians and 3,096,600 by other non-nationals.[69] This is not a new phenomenon in Paraguay as millions of hectares were sold to foreign enterprises after the Triple Alliance war. However, there is a recent increment that is related to the land acquisition processes that occurred worldwide in the first decade of this century.

Both land concentration and foreignization are pushed by the extractivist development model based on agribusiness implemented by the Paraguayan government. This model focuses on the production of soy and beef. The soy production boomed during the last 15 years: the land used for soy increased from 1,200,000 hectares in 2000 to 3,264,000 hectares in 2016.[70] Soy production rapidly expanded to almost all the agricultural land available in the Eastern region of Paraguay. In 2015, around 37% of the soy production[vii] was directly exported to the European Union.[71] Despite the on-going research conducted to develop "improved" genetically modified seeds adapted to the Paraguayan Chaco region, almost 100% of the Paraguayan soy is produced in the Eastern region. In 2015, Paraguay was the 4th largest world soy exporter (after USA, Brazil and Argentina). The last Agricultural

[vi] *The Gini coefficient measures, in this case, the land unequal distribution. 0 represents the perfect equality and 1 the maximal inequality.*

[vii] *This is not related with the vegan or vegetarian consume of soy. Paraguayan soy is genetically modified and is mainly used to feed animals.*

Survey clearly shows how soy production, land concentration and foreignization interrelate. 64% of the land for soy production was owned by foreigners.[72] The same year, 64% of the soy was produced in agricultural properties with 500 or more hectares.[73]

Paraguay is also among the six largest world beef exporters, following Argentina.[74] The main actors of the cattle ranches' expansion in the Chaco region are the Mennonites colonies, the Brazilian ranchers and the Uruguayans ranchers. This last group bought 12% of the land used for cattle production in Paraguay, in the past decade.[75] The total number of cattle in the Paraguayan Chaco increased 7.3% from 2013 to 2014.[76]

Land concentration and foreignization are interrelated with incredibly high deforestation rates. According to deforestation monitoring reports, only in the Paraguayan Chaco 268,437, 236,869, 287,435, 285,526, 189,472 hectares were deforested in 2012, 2013, 2014, 2015 and 2016 respectively.[77] The highest deforestation rate per day was reached in 2012 and 2013, with a world record of 2,000 hectares per day.[78] Hansen et al. revealed that the deforestation dynamics in the Chaco woodlands of Paraguay, Argentina and Bolivia represented one of the highest rate of tropical forest loss.[79]

The soy boom and land acquisition process changed the ranching style in the Chaco region, in general, and in the Angaité area, in particular. Global political and economic dynamics affected, simultaneously but in different ways, other indigenous groups of the Paraguayan Chaco. Blaser, for example, mentions that in 1996, "*as the Paraguayan transition advanced and the integration of Paraguay into Mercosur deepened (…) new economic enterprises began to emerge.*"[80] In that case, Brazilian entrepreneurs began to buy and clear-cut the forest to establish cattle ranches. This new cattle ranch management style affected the local indigenous communities because the new owners were far less tolerant than the older ones and increasingly restricted the access to hunt on their properties.[81]

Due to the lack of a functional national cadastre, it is not easy to access basic information about private property ownership in Paraguay. Information about economic activities conducted and natural resource use in private properties is even harder to find. Despite submitting several official requests to the corresponding governmental offices, I was refused any official information. In order to describe what was going on in the ranches located in La Patria area, I enquired locally regarding the name of each ranch and its owner. Using these first hints, I navigated the internet searching for information. Using what I found in the internet, I was able to partially fill this gap. I used similar strategies in previous studies as did other recognised national researchers.[82]

Figure 8. Forest cover loss 2000-2014 in the Paraguayan Chaco region.

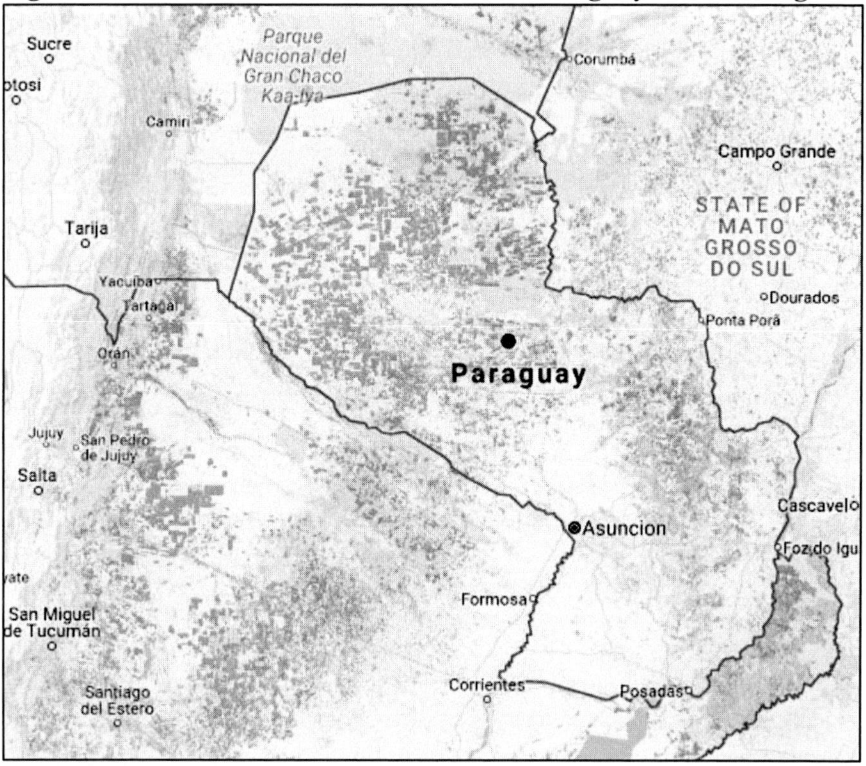

Grey areas: Forest cover loss. Black dot: La Patria colony. (M. C. Hansen et al., 2013)

The only official document, available for some of the ranches, is the *Relatorio de Impacto Ambiental*[viii] (RIMA). Regarding natural resource use and management, cattle breeding activities are subjected to Law 294/96 and some other related decrees: 954/13 and 453/13. The RIMA presented by the cattle ranches to the SEAM in order to get the environmental permits is one important source to describe how the natural resources are managed on the ranches. Though these RIMAs should be accessible to the public, after many

[viii] *Environmental Impact Report*

attempts I only managed to access the RIMA of some ranches located in the area of La Patria.

Only one of the above-mentioned law and decrees references indigenous communities. The decree 453/13 states that if the project impacts indigenous communities directly, the SEAM has to inform the community and organize a public hearing. However, "direct impact" is not clearly defined. As the following examples show, the environmental consultants of the RIMA usually delimit the direct impacts to the property limits and indirect impacts to an extension of 1,000 meters from the latter.

Some of the ranches located in or close to the area used by the Angaité of La Patria ("La Patria territory") still have an extensive pattern of resource use. Others are using their lands at its maximum carrying capacity. Environmental consultants consider that the carrying capacity in this area is around 0.8 cattle heads per hectare.[83] Estancia Riacho González, owned by Dictator Stroessner's close collaborator Bader Rachid Lichi, is a good example of a ranch in a middle point. This ranch of 14,500 hectares, located directly North of La Patria, has approximately 7,000 heads of cattle. The RIMA of this ranch was not accessible but local informants provided some information on how it functions. The ranch has 10 permanent workers, out of which 5 are Angaité from La Patria. Considering that the law establishes that a minimum of 25% of the property surface has to be conserved as a forest reserve (3,625 ha in this case), the maximum number of heads of cattle should be 8,700. This means that *estancia Riacho González* runs below its maximum potential. However, consultants calculate the carrying capacity using an average weight of 400 Kg per adult animal, under favourable climatic conditions, applying a rotation system that allows the pasture to regenerate.[84] In other words, under current weather conditions of extreme droughts and wet seasons, the maximum is certainly lower than 0.8 cattle heads per hectare.

The development or expansion of cattle ranches involves the following practices: forest clearance, burning, building and opening of roads; implementation of pastures; and establishment of water reservoirs. All these practices were approved in 2015 as exemplified by *estancia 14 de Mayo*. This ranch of 12,980 hectares is owned by the Paraguayan Orue Ayala family.[85] This family retains property in this area since the 1970s. Despite being neighbours of La Patria this community is not mentioned in the RIMA. The possible negative impacts of clearing 2,266 hectares of native forest on La Patria population is not considered. In fact, the RIMA describes the region as an "*exclusive cattle production ecoregion*."[86] Including the pastures to be implanted, the ranch will have 3,278 hectares of pastures, to be used at it maximum carrying capacity (in this case 0,7 animal/hectare). This ranch, was authorized to maintain only 2,173 hectares of forest reserve (only 16%, this is, below the 25% required by law) because it represented 40.5% "*of the forest that originally existed in the property.*"[87] In order to mitigate the negative impacts of deforestation, the environmental consultant that elaborated the RIMA suggested the selective clearance of trees with less than 20 cm diameter at breast height. However, the clearance was conducted by using two "D6" CAT bulldozers. These activities were planned for a period of 2 years. Among the positive impacts RIMA mentions the production of food: "*Food is a basic human need; this project plans to produce meat for human consumption, which demand is constantly increasing due mainly to the population growth.*"[88] A supposed positive impact is the creation of job opportunities. However, the number of jobs created by this kind of ranches are considerably low: this 12,980 hectares' ranch hires a total of 10 permanent workers (1 workplace every 1,298 hectares) and in some seasons temporal workers are hired (Lampert 2015).

Some of the RIMA are very detailed when it comes to listing the negative impacts. Directly after this long list there is a series of recommendations that state both the positive impacts and the lack of alternatives that justify the drawbacks. Some of the negative impacts mentioned in the *Estancia Laguna Rey'* RIMA, a ranch of 17,152 hectares owned by COBANER S.A, were: impacts on the natural landscape; emission of carbon dioxide (CO2) and carbon monoxide (CO) related to the incremented cattle population; loss of natural forest cover and biodiversity; migration of fauna species due to habitat destruction and hunting; soil compaction, salinization and soil loss due to hydric and eolic erosion; competition of cattle and wild fauna for food sources and vital space and impacts on the soil geomorphology; and reduction of it infiltration capacity.[89] This RIMA defines the "direct influence area" as the ranch property and "Indirect influence areas" as the area up to 1,000 meters from the property.[90] This RIMA omitted mention of La Patria colony.

A fourth RIMA shows how the existence of indigenous communities and "La Patria territory" is treated when mentioned in the RIMA. The RIMA of a 6,499 hectares' ranch owned by "Guillermo Enciso e Hijos Ganadera, Forestal e Inmobiliaria S.A", located a few kilometres East of La Patria, explicitly states that the forest to be cleared is used by indigenous people. It states: "*There are people who live of the forest products, like most of the indigenous people, which constitute a sector that would be seriously affected*"[91]. Following this, the document states: "*there would be changes, but the important thing is that these changes should be the least traumatic possible for all the actors.*"[92] No mitigation measures or any particular way of reducing the possible negative impact to the Angaité are suggested. Indigenous hunters are only mentioned once more, when they are blamed for the fires that threaten the forests.[93]

The four RIMAs analysed here are not an exception. They are totally biased and favour ranchers over indigenous communities. This is repeated in other RIMAs of ranches located nearby indigenous communities in Paraguay. Recognised environmentalists claim that the strategic location of the ranches is not generally considered, specifically whether there is an indigenous territory within the property.[ix] Even worse, the RIMA are normally desktop studies (consultants have usually not been to the field), so they rely on general and outdated literature.

The accumulative impact of independent projects is not considered by the SEAM to require the issuing of environmental permits. Every RIMA considers the possible impact of the deforestation of a certain amount of land. However, the SEAM is not analysing the overall cumulative impacts of the regional deforestation of thousands of hectares. This was, at least formally, considered by the environmental authority of the Argentinian Chaco region since 2009. Finally, the RIMA are not mentioning the increased occurrence of floods in the area. Recent research conducted by the National Institute for Agricultural Technology from Argentina proved that the native Chaco forest absorbs up to 300 millimetres of water per hour, while an implanted pasture with cattle only absorbs 100 millimetres.[94] The scholars who conducted this research argue that the "*hydric excess are not due the lack of infrastructure or the rain excess, but due to the deforestation and implantation of pastures in the last two decades.*"[95]

This was considering this situation that the UN Special Rapporteur on the rights of indigenous peoples expressed her concern regarding violations of indigenous peoples' land rights by public institutions. Mentioning, in particular, "*the issuance of environmental permits by the Secretariat for the Environments that do not conform to Paraguayan law or international standards relating to the rights of indigenous peoples.*"[96]

[ix] *see Elías 2008 in Marcos Glauser, 2009*

3.3.3 Land use of the Angaité of La Patria in 2015

How compatible were these new practices with the Angaité land use practices? In this section, I explore how the Angaité of la Patria changed their land use patterns in order to respond to the recent territorial dispossession caused by the deforestation and access loss provoked by the new intensive ranching practices. During my fieldwork, I explored which areas were used for *marisca* (hunting, fishing and gathering) during 2015 using a participatory mapping methodology. My participatory mapping in La Patria can be understood as part of wider experiences generally called "indigenous mapping". These experiences record and reveal the indigenous territories that exist in spite of and counter to "official" maps produced by states.[97] However, I have mapped the land use of La Patria instead of the Angaité territory as an ethnic group. Unlike other indigenous mapping exercises, I did not try to demonstrate any historical continuity with a pre-colonial past or to map a definitive territory used by this community. In fact, I show that areas and resources used did change over time. In order to avoid any future misuse of the maps I produced, I need to underline that these maps do not represent the way things are. Rather, these maps are a self-consciously fashioned representations that result from a particular methodology and that must conform to certain conventions (Bryan 2009).

There are few official maps in Paraguay for the 493 indigenous communities that officially exist.[x] Basically, one of the few sources is the Atlas of Indigenous Communities[98] published by the Indigenous Population Census results. These maps show where indigenous settlements are located. They do not include the real borders around each communal land property. These maps are misleading because they show (in light grey buffer) the "approximate area of the community". One might think that each of these villages is a separate community with its own land property, which is not the case.

The land use of indigenous communities is misleading because the map does not contain any information on use. The map does not represent, it renders invisible, the overlapped territory I previously described as "La Patria territory. This implies that the consultants in charge of the RIMA do not

[x] *According to the third Indigenous Population Census, there are 493 communities, 218 villages or neighbourhoods, totalizing 711 indigenous settlements nationwide.*

have the information required to consider the multiple uses of a particular area at their disposal.

Figure 9. Representation of La Patria colony and its villages in the Atlas of Indigenous Communities (DGEEC, 2004)

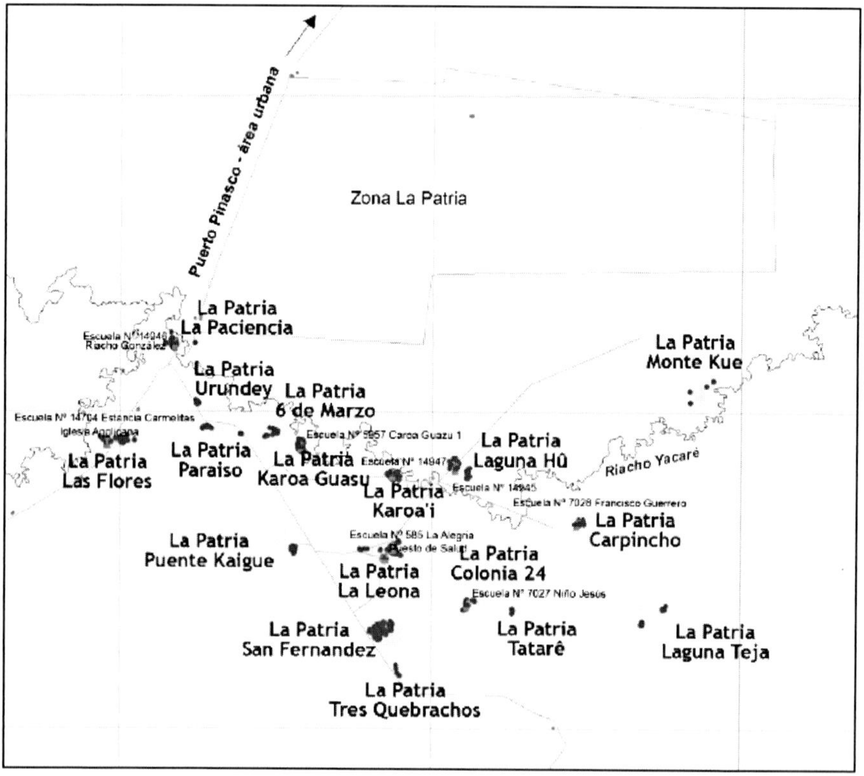

During the participatory mapping of the land use patterns and livelihood practices conducted outside of the colony I did not ask the people to draw maps on paper. Most drawings were used to map the land use inside the colony. Instead, I relied on the information which leaders and members of each village had given me when I interviewed them. Maps were drawn on the ground during these interviews. These maps depicted the ranches where the Angaité work and hunt as well as other routes used to hunt inside cattle ranches. As Leake observed in 1995, people invariably erased these maps once they finished drawing them. People's spatial awareness appeared to be restricted to the areas they and their fellow villagers used. So, my land use map is the result of information gathered about the areas used by each village. To stimulate my interviews, I used an old regional map provided by Leake and probably made by the IPC. Interestingly, elderly people were very excited to listen to the names of old ranches and places. However, they didn't show any particular interest in the map itself, remaining distant and simply asking me to read aloud one place after the other. Every name triggered memories and were immediately related with people's histories. As Tytelman suggested, naming places should be considered an important "place making" practice.[99]

As mentioned before, the territory used by the Angaité in 1995 has been under an intense process of deforestation since then. Figure 10 shows, with a dotted polygon, the area used by the Angaité in 1995. The area used by the Angaité to hunt in 1995 was recalculated based on more accurate information that was not available when Leake conducted his research. Based on his description I identified both the land the ranches that existed at that time occupied and the areas the Angaité used for hunting. The resulting map is quite similar to the map produced by Leake, but the total land area is considerably lower. Instead of the 240,000 hectares mentioned by Leake, my recalculation indicates that the extension used in 1995 was of approximately 209,000 hectares.

Figure 10 also shows the territory used by the Angaité of La Patria in 2015. In total, it covers an area of 174,000 hectares. This means that it decreased 35,000 hectares, a 16.7% reduction. The territory used was reduced in its Southern, Western and Eastern extremes (ranch 10, 15, 16 and 20 respectively). However, it extended a few kilometres to the North (ranch 3). A considerable amount of people mentioned that they currently hunt in the cattle ranches where they work in the Mennonites colonies of Filadelfia and Loma Plata. Some are also working in *estancia Lucero*, close to Pozo Colorado, and in a ranch situated between Loma Plata and *Puerto Casado* cities. This indicates that the pressure on the local environment was reduced by work migration to other places. These places are not included in the map because they are too far away.

The black areas in Figure 10 represent the areas affected by deforestation for pastures in cattle ranches in the period from 2000 to 2014. This period is used due to the lack of satellite images from 1995. In order to avoid misunderstandings, the reader has to keep in mind that some areas were deforested before the year 2000. However, most of the deforestation that affected this region occurred after the year 2000, and as such, the phenomenon is well documented. According to my GIS analysis the total tree cover loss from 2000-2014, inside the area used by the Angaité in 1995, was of approximately 27,757 hectares.

Figure 10. Deforestation of Patria territory. Areas used for hunting, gathering and fishing in 1995 and 2015. (Forest loss data source: M. C. Hansen et al., 2013)

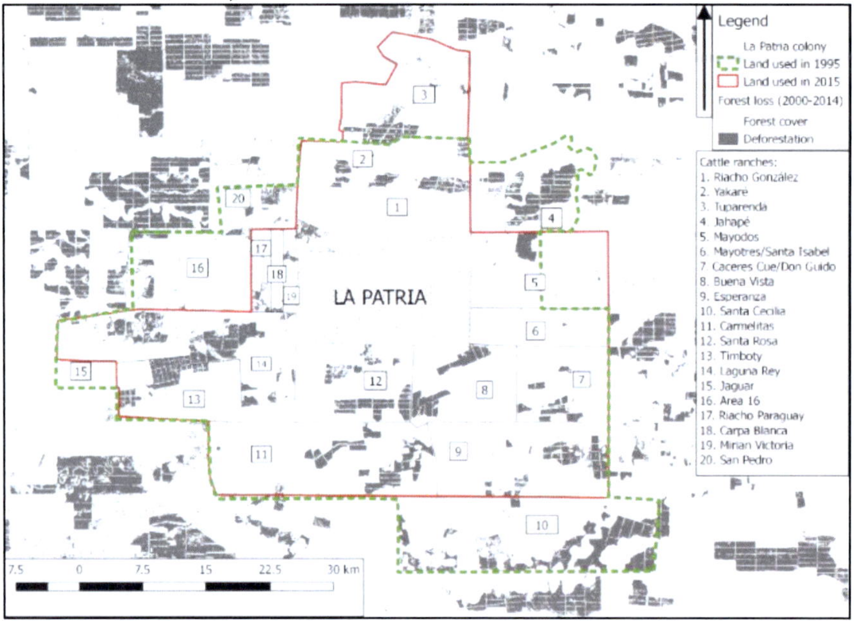

At village level, the land use change is more comprehensive. Some villages have stopped hunting in some ranches and started to do it in others. Some ranches were abandoned by all the villages. Other ranches were abandoned by one village and used by another. Table 8 tracks these changes (see Appendix 1). A comparison between ranches is not possible due to a series of factors. Leake lacked information about the limits of the ranches. The limits drawn on his maps are approximations and do not always correspond to current ranches' limits. In addition, the limits of some ranches changed with time as the property was divided or united. However, Leake's map allowed me to trace some comparisons.

There were 19 different ranches used by the 10 villages that existed in 1995. 13 of these ranches were still used in 2015. I tried to find out whether the villages were still using the same ranches. I found thirteen ranches that were continuously used, they were: *Karova Guasú* village with *Carmelitas* and *Mayodos* ranches; *Karoa'í* village with *Carmelitas* and *Laguna Rey* ranches; *Carpincho* village with *Mayodos* and *Mayotres* ranches; Comunidad 24 village with *Buena Vista* and *Esperanza* ranches; *San Fernández* village with *Santa Rosa* ranch; *Las Flores* village with *Mirian Victoria* ranch; *Monte Kué* village with *Mayodos* and *Riacho González* ranches; and finally, *Tatarẽ* village with *Buena Vista* ranch. However, due to the creation of 8 new villages in La Patria it was not possible to track the existence of other continuous relations. Let me explain this with an example, villagers from *San Fernández* hunted in *Carmelitas* ranch in 1995 and apparently the relation finished because they did not mention this ranch in 2015. However, villagers from *Puente Kaigue* that were living in *San Fernández* village in 1995 still hunted in *Carmelitas* ranch in 2015.

The most popular ranches in 2015 were *Carmelitas* (used by 7 villages), *Yakare* (used by 5 villages), *Buena Vista* (used by 3 villages), *Mirian Victoria* (used by 3 villages), *Mayodos* (used by 3 villages), and *Laguna Rey* (used by 3 villages). Six of the nineteen ranches used in 1995 were in disuse in 2015, they were: *Carpa Blanca, Sta. Cecilia, Jaguar, Area 16, San Pedro and Jahapé.*

Some ranches were more deforested than others in this period. One might be tempted to relate the abandonment of certain ranches with deforestation. For example, "Sta. Cecilia" ranch lost a considerable amount of it forest cover and was indeed abandoned. The same goes for *Jahapé* ranch. Not all abandoned ranches were deforested, such as ranch "Area 16". This seems like a great contradiction, yet it is not surprising. Almost all "popular ranches" lost forest cover in recent years or were cutting their forests by 2015. This means that these ranches were hiring more people than the others. Therefore, the Angaité were gaining access to those areas in 2015. As contradictory as it may sound, deforestation is associated with employment opportunities and to temporal access to the forest and wildlife resources on those particular ranches.

In short, the total area used by the Angaité of La Patria for hunting, fishing and gathering slightly decreased over the past 20 years. Work migration to ranches located in other regions of the Chaco allowed them to hunt in areas which they did not use in 1995. The way deforestation affected their livelihood strategies is not reflected in this spatial analysis but will be considered in the next chapter. In the next section of this chapter, I focus on additional information the Angaité provided regarding the complex territorial dispossession process they experienced over the past 20 years.

3.4 What else did the Angaité say about the recent land use changes?

When I talked to the Angaité about deforestation and the changes that occurred over the past 20 years they immediately related my research interest to that of a tall anthropologist they nicknamed *ñandú.* I took a while to realise that they were talking about Leake. The Angaité usually do not refer to specific times or years when reporting events. They told me how they lived when *ñandú* lived among them. As usual, they didn't refer to a specific year, they just mention that this or that event happened in "*ñandú tiempo*" (*ñandu*-time). Besides his accurate land use information, the mention of Leake's nickname became a useful time reference.

Most of the people to which I showed Leake's maps were excited to see the results of research conducted 20 years earlier.[xi] Some leaders immediately understood what those maps were communicating and commented that it would be a good idea to show these maps to journalists from the national media "*so that the people would see how the indigenous people use the land, as it can be clearly seen in the maps. We use the territories that we know because we have lived there before. We have to go back there because there are not too many animals left inside La Patria. But the ranches are cutting all our forest, our supermarket*".[xii] It is interesting to note that this leader referred to the Angaité as "indigenous people", assuming this as an identity category. He referred to the forest located inside privately owned properties as "our forest", this implies a sense of ownership or use rights over it. In a clear translation effort, he followed by describing this forest as "our supermarket" to imply that it is there where they get what they need.

Leaders described their dependence on forest resources to evidence that their way of relating to and using the forest was distinct: "*the ranchers are cutting the forest, we don't, because there are many important things in the forest, honey,*

[xi] *Only some leaders that had worked with Leake knew these maps.*

[xii] *"…para que la gente sepa cómo los indígenas usan la tierra, que ahí se ve muy bien. La gente va y accede a los territorios que conoce luego porque ahí vivía y ahí nomás luego la gente tiene que volver a ir ahora, que en La Patria hay pocos animales. Ganadero kuéra oitypáta ñande ka'aguy ñande mercado." Mr. Santiago, June 2015, my emphasis*

animals…. And deforestation brings many problems".[xiii] The unexpected abundance of certain wild animals in the colony was related to deforestation in nearby ranches. As exemplified by estancia *Yakaré*, where thousands of hectares were cleared in 2015. I soon realised that my research interested them because they were conscious of and concerned by the magnitude of the changes in their region and in the rest of the Chaco. One of the first villagers from *Karoa'í* that I interviewed mentioned that the leaders were considering reaching out to the owners and managers of the neighbouring ranches before "*…they deforest too much their properties, the areas limiting with La Patria colony. I saw in Boquerón how ranches neighbouring indigenous communities are cutting everything (…) Mennonites are buying land there. Also in Mariscal Estigarribia area, Pycasu (indigenous community) and Picada 500 the same is happening, the Argentinians and Mennonites are buying land there…*".[xiv]

In my first series of interviews I asked the Angaité directly what had changed over the past 20 years. The answers I got are coherent with what I later found through the participatory mapping. Firstly, leaders and villagers said that there was more forest in the neighbouring areas in 1995. Secondly, they observed that people started to work in ranches located far away because the local ranchers were paying very little, usually just providing provisions. Finally, regarding land used, they said that the limits of the territory they currently used were *estancia Algarrobo* and *Laguna Rey* (to the West), *estancia Tuparendá* and *Yakaré* (to the North), *Colonia Ceibo* (to the East) and *estancia Carmelitas* and *Esperanza* (to the South). They mentioned that other communities - who also work on the ranches - hunt, gather and fish beyond those borders.

One leader of *Karoa'í* village described the process of enclosure that occurred in the past 20 years as follows: "*As our surroundings were fenced, the ranchers started to restrict the access and we were not allowed to hunt fish and gather*

[xiii] "*…ganadero oitypa la ka'aguy, ápe nahániri, porque en el monte hay muchas cosas importantes: miel, animales… ha el desmonte trae heta perjuicio.*" Mr. Daniel, June 2015

[xiv] "*iñimportante ante que la ko´ãva ningo odeforestaiterei ikámpo, kóva ko zona de La Patria che la ahecha. Che ahecha pe zona de Boquerón vecino de la comunidad indígena, upépe perõ... (...) menno oike ojogua, ivesíno voi, omoperõ... Ha pe zona de Mariscal amo gotyo, pe pycasu, oho ojogua umi menno avei, sa'íma ohecha... Mariscal jaikére picada 500 jaike jaha upe zona de frontera, péva péicha avei, umi kurepi, umívante la oreko umi... ojoparapáma, oĩ la kurepi.*" Mr. Mateo, June 2015

anymore. Before that, people used to go one or two leagues[xv] *and knew where forest and fruits were abundant. When everything started to be deforested,* *it seems that the wild fruits season never reached again and the shortage started.*"[xvi]

The loss of access was mentioned in various interviews. It was around the year 2000 when admittance into most neighbouring properties started to be restricted. However, this process took place at different times and in particular ways in each village. Interviewed villagers from *La Paciencia* said that in 1995 "*you could move around, everything was free. Now it is forbidden to enter or to hunt, gather and fish*".[xvii] Workers from *estancia Riacho Gonzalez*, the closest ranch to *La Paciencia* village, said that no restriction were imposed on indigenous hunters. This version was opposed by the villagers of *La Paciencia.* They claimed that only people that maintained "good relations" with the ranch's care taker by letting him know every time they wanted to enter were allowed to hunt there. The same situation applies to *Monte Kué* village, also situated close to this ranch. Villagers from these two villages mentioned the importance of the resources located in the south-east part of this ranch, where game and honey are abundant. Access for villagers from other villages of La Patria was not permitted.

A similar situation was experienced by *Carpincho* and *Monte Kué* villages since 2002. They stopped hunting, gathering and fishing inside *estancia Mayodos* that year because a new ranch administrator was hired. He imposed new rules, including a requirement that the Angaité ask for permission before entering. The new administration's rules and attitude were said to be related to a fear of cattle rustlers. A few years later, in 2006, the same rules were applied in *estancia Mayotres* and *Buena Vista*. "*Now you can only enter if you ask for permission and if the care taker knows you. But you are not allowed to enter ridding a donkey. And if you want to get some honey, you cannot go so far because you will have to carry everything.*"[xviii] Since then, some villagers from *Tatarē* and *Colonia 24* still

[xv] *A league is a unit of length. It originally referred to the distance an individual could walk in an hour (4-7 km)*

[xvi] *"A medida que se fue alambrando a nuestro alrededor, se fue prohibiendo y ya no pudimos más entrar a mariscar. Antes la gente se iba uno o dos leguas y sabía donde abundan las frutas y monte. Después, cuando se empezó a deforester más, es como que no llegó más el tiempo de las frutas del monte y todo empezó a escasear." Mr. Santiago, June 2015, my emphasis*

[xvii] *"…antes podías moverte, todo era libre, ahora está prohibido entrar a mariscar." Mr. Antonio, November 2015*

[xviii] *"Ahora solo se puede entrar si uno pide permiso y tiene que conocer al encargado, no se puede entrar en burro y si uno quiere sacar miel, no se puede ir lejos, ni puede traer tampoco tanta carga." Mr. Diego, September 2015, my emphasis*

rode to *estancia Esperanza*, close to *Colonia Ceibo*, where they entered without asking permission because there were no cows in that paddock yet.

A slightly different way of losing access is exemplified by the relationship between *Tres Quebrachos* village and *estancia Santa Rosa*. This ranch deforested its North-East section in 2013. This section is directly next to La Patria's southern border and was used by villagers till then. In this case, there were no access restriction, but the deforestation destroyed the resources and the animals moved somewhere else.

The following statement reflects a sense of ownership over lands located beyond the colony limits which were lost when ranchers built fences. In 2006, eleven years after Leake's research, Agapito Navarro stated:

> *"These foods are now not easy to find. Because we own not much land, our territory was fenced and we cannot go around looking for food. (…) Cattle ranchers ended up owning everything and have fenced the land. What else could we do now, if in our colony's land there is not enough (food) for everybody? We are neither able to get honey nor wild animals. There are few deer and armadillos. We cannot go around looking because they forbid us to enter in private properties. If we trespass we risk of being killed."* [xix]

Access restrictions were considered by the Angaité as one expression of the *Paraguayans'* change of attitude towards indigenous people. One fifty-year-old woman said: *"That is what I am saying, that is what I said. Now the Paraguayans are mean towards the indigenous people. It was not like that in the past. Now there is no permission any more to enter the estancias. Sometimes they give permission. Sometimes they don't"*.[xx] A similar change of attitude was experienced when the Paraguayans first reached Angaité territory. Villagra mentions that an initial cooperation between the settlers and the Angaité was quickly replaced by abuse and mistreatment. The latter is evidenced by the Agaité's forced displacement and the settlers' control over land.[100] The Angaité of La Patria refer to an improvement of Paraguayan-indigenous relations in the last decades of the previous century but a change of attitude at the beginning of this one.

As mentioned before, visiting relatives on cattle ranches was increasingly

[xix] *Mr. Agapito Navarro in Grupo Sunu 2007, my emphasis*

[xx] *"Chéko péva la ha'e, ha'e voi che, ko'ãga paraguayo kuéra iñaña la indígenandi. Yma nahániri. Ko'ãga ndaiporivéima permiso emariska estánciape. Algunas veces ome'ẽ, algunas veces nome'ẽi…"*

restricted in the past years. This threatens one of the Angaité's strategies to access private properties. The Angaité relate these new rules to new ranch owners, administrators, care taker *(encargados)* or foremen (*capataces)*. The new rules are not associated to the more intensive ranching style but to newcomers with a different attitude. The Angaité from La Patria have slowly built personal relations with ranch workers and that is why a staff removal means a big step back for them. Over the past years, the staff members of most ranches were renewed. In some cases, the renewal ensued from the sale of the ranch. In other cases, the staff was replaced because the ranchers distrusted them. The Angaité describe the time before this renewal as follows, "*the ranchers were not mean yet and we were tranquil*."[xxi]

Angaité perception of deforestation in cattle ranches must be considered in the context of access loss. The words of a leader I interviewed beautifully encapsulate this observation:

> "*…we have a double view. For one side, we don't like a deforested forest, because it is not accessible any more, it is privatized. The forest has no owner. Is everybody's forest. We are not allowed to enter the pastures. So, more deforestation means fewer possibilities to hunt and gather. On the other hand, it creates job opportunities.*"[xxii]

Deforestation, in this context, creates temporal opportunities to access the forest reserves in the ranches.

The Angaité *de jure* have the right to access any private property for livelihood related activities but *de facto* they need to negotiate this access with the ranch staff. The Angaité know their legal rights. Mr. Ignacio, leader of *Tres Quebrachos* village said that

> "*according to the Law 904, according to our culture, we don't need to ask permission, we are allowed to go wherever we want to. This is what the law says. But sometimes the whites, the ranchers, don't know our law (referring to 904), and that is sometimes confronting us, because they don't understand our law (…). We don't want to steal anything, we only want to hunt, gather and fish. The wild animals are ours anyway. We have to use that forest*".[xxiii]

[xxi] *"patronkuéra ne'ĩra gueteri iñañaha, ha roime kuri tranquilo" Mr. Diego, September 2015, my emphasis*

[xxii] *Mr. Santiago, June 2015, my emphasis*

[xxiii] *"asegún la ley, la ley 904, asegún ore reko, ndaikatúi rojerure, rohosehápe roho va'erã, asegún la léipe oĩ. Pero pe sapy'ánte he'i la blanco, la ganadero kuéra, ndoikuaái la ore léi, ha péva sapy'ánte ha'etéko voi ore choka jey, porque ha'ekuéra ndopillái la ore léi, oimehamehâme ikatu roho roike. Ndaha'éi roho haguã*

Access prohibition is not perceived by the Angaité of La Patria as a failure to respect existing laws but as a matter of "not knowing" or "not understanding". This is why the art of convincing and negotiating access is considered an important skill. This skill is cultivated in childhood. A villager explains, "*my father taught me how to treat the whites, to the care takers, in order to ask permission to hunt, gather and fish.*"[xxiv] Others said that they had acquired this skill by living their childhood in cattle ranchers or being raised by *paraguayos.* In two cases, villagers mentioned that they had learned how to treat the whites by being beaten by their *patrones* in some ranches. Actually, a considerable number of current leaders lived at least a few years in ranches.

In a meeting organised by the leaders of La Patria in *La Leona* village, to discuss increasing access loss, the main argument expressed to legitimise their right to access was not the law but their own historical dwelling on those ranches. In a long intervention, worth quoting, Mr. Santiago, leader of *Karoa'í* village, literally said:

> "…*there are two things that link our people with our territory: work in cattle ranches and marisca* (*hunting, gathering and fishing*). (…). *People from La Patria go and wander around and then talk to their father and grandfather, and they have the history. Those who go around do not know why they are going there once and once again, why they always work in those ranches. It is because their parents and grandparents used to live there before. We need to strengthen that memory* (…). *We should publish this through the newspapers and television. By doing so other people would know who are the indigenous people that live in this area of La Patria, where they live, where they work, where do they get their food from. Because sometimes the food is finished here in the colony, and people go to the forest, to their old settlements. They remember their old settlements. They used to live there. And they have to go there again to get wild animals and honey.* (…) *They (people from La Patria) have to go there because their parent also got that from those places, and as they were kids they (people from*

romonda, solamente roho haguã de maríska umíva. Vícho ka'aguy ore mba'e voi. Ore roiporu va'erã pe ka'aguy." Mr. Ignacio, June 2015, my emphasis

[xxiv] *Mr. Teodoro, October 2015*

> <u>*La Patria) followed their parents to these places.*</u> *They know where to go. That is why they know where to look for food now.*"[xxv]

This quotation speaks for itself. The ranches might have privatized the land but it is still their territory and their ancestral relation to it remains alive. Through work and *marisca* they visit and communicate with their territory. They have been continuously present there for at least a couple of generations and maybe even longer. The Angaité try to get ranchers and others to recognize who they are, where they live and what they eat because they feel invisible. Knowledge of their territory, which includes much more than "natural resources", is acquired first hand by walking or second hand from their parents' teachings.

3.5 Conclusion

I started this chapter exploring, in section 3.2, the Angaité and the ranchers' cosmography. I argued that for the Angaité, people, non-humans and other "not so human" form part of a wider "social collective". For them, "nature" constitutes part of a complex network of social interactions in which man is no more than one actor amongst many. As in other 'organic nature regimes' rituals and practices embedded in social relations establish continuity and links. For the Angaité, the *choqueo*, in which the role of shamans as singers is central, still constitutes an important cosmographic practice. This ritual and the shaman's powers were used to protect the communities and provide with material benefits. However, interactions with non-human beings are not confined to rituals or shamans alone. The *marisca* (hunting, gathering and fishing), a central livelihood activity, involves a set of practices through which the spiritual and the material as well as the human and non-human are amalgamated. These rituals, shamanism and the *marisca* are not static. As

[xxv] *"…dos cosas ovinkuláva la territóriondi la géntepe: chánga ha maríska. Umi dos cosas ko'ãga peve ovinkuláva. (…) Ko'ápegua oho orrekorre, ha upéi oñomongeta avei itúva, ijabuélo kuérandi, ha hikuái oguereko la historia. La ohóva oheka ndoikuaái mba'érepa la cada viaje oho upépe omba'apo jepi. Porque upépe voi itúva kuéra umíva oĩ va'ekue. Entonces romombareteve upéva pe, memoria... (…) Péva ikatu ñembojekuaa diário o télepe. Upépe ikatu ohecha hikuái mávapa la indígena pe zona de La Patria oĩva, moõ moõpa oiko, moõpa omba'apo, moõpa onohẽ la alimento kuéra umíva. Porque la comunidad ningo... ko'ápe ningo sapy'ánte opa, ohopa ka'aguýpe, hekohague voi. Omandua hekohaguépe. Upépe oiko ha'e. Upépe oikuaa moõpa oike pe ka'aguýpe, moõpa oheka la animal, vícho ka'aguy umíva. Upépe voi oiko. Ha upépe jeýnte oho va'erã ogueru haguã la vícho ka'aguy o eíra. (…) Upepente oho va'erã. Porque itúva ohóne upépe raka'e, ha mitã'i umíva oho hapykuéri, ha oikuaa moõpa oho. Ha'e ojeporeka jeýta oikuaahápe mante oho va'erã." Mr. Santiago, June 2015, my emphasis*

shown by Villagra, the choqueo experienced important transformations since the proliferation of cattle ranches and the subsequent dispossession of their territory.

Then, I presented the cosmography embodied in the cattle ranches as central to the "development" cosmography deployed in the Angaité area since the early 1900s. I argued that this cosmography experienced important changes since the beginning of 1990 and that these changes were evidenced in the new intensive ranching style.

The initial hints about the Angaité cosmography, mainly provided by Villagra's and Bonifacio's work, my own fieldwork and some parallelism with Tola's ethnography, shed some light on

The second dimension of "La Patria territory", the fabric of social relations between beings that constitute a territory, is explicated by using my own fieldwork, references to Angaité cosmography - mainly provided by Villagra's and Bonifacio's work - and parallelisms with Tola's ethnography. For the Angaité, the multidimensional cosmos is inhabited by human beings, animals, plants and many other types of beings, called *apmovana* (auxiliary spirits) that live in the different levels of the cosmos.[101] Humans relate with these beings that can help their human partners to master the powers controlled by them. This is why Tytelman definitions of place as *"an assemblage of enduring (but changing) patterns produced by the interlinked practices of people and non-human persons"*[102] proved useful to understanding the Angaité cosmography. Angaité shamans' capacity to communicate with these auxiliary spirits was used to protect their communities and provide with food. Thus, it is clear that the "political domain" was not restricted to the human social domain. What I mean to say is that the "social collective" is not defined by a clear distinction of "nature" and "society" or "beings" and "things".

The Angaité's cosmography and the rancher's "development" cosmography are not isolated worlds but have been related since the colonization of the Angaité territory. The ideas and practices that constitute each of them were, at least based on Leake's descriptions, not completely contradictory. So, the set of practices involved in making a productive ranch, in 1995, were limiting but not impeding the continuation of the practices that constituted "La Patria territory".

In section 3.3 I focused on recent territorial dispossession and land use in 2015. I explored the extent to which the ranchers' natural resources use practices, not only allowed but promoted by the Paraguayan state, affected the continuity of the Angaité land use practices in the past 20 years. I first briefly presented the land use patterns of the Angaité of La Patria in 1995, based on Leake's results. Secondly, I described the main changes that affected "La Patria territory" since 1995. My focus was on the new intensive resource use practices associated with the new ranching style. I showed that this new ranching style is related to the process of land concentration and foreignization caused by the expansion of agribusiness in Paraguay. The effects of the development of new ranches and expansion of old ones were evidenced in deforestation rates of 2000 hectares per day in the Paraguayan Chaco in 2012. In the Angaité area, the new ranching style implied not only the deforestation of forest resources used by the Angaité but also increasing restrictions in the access to private properties to conduct their livelihood practices. By analysing the RIMA of cattle ranches that overlap with or are close to "La Patria territory", I showed that in practice the Angaité territorial rights and "La Patria territory" are not respected nor recognized by the Paraguayan state. In other words, the Paraguayan state promoted and allowed practices that were partially incompatible with the Angaité land use practices. Then I presented the Angaité land use practices outside La Patria colony in 2015. I found that the territory they used outside their colony decreased 35,000 hectares (16.7%) from 1995 to 2015. In this period, some villages stopped hunting, fishing and gathering in some ranches and started to so in others. Out of the 19 different ranches used in 1995, 13 were still used in 2015. Many villages were still hunting in the same ranches where they hunted in 1995. The following strategy is in use since 1995: village members undertake paid jobs in ranches to ensure access to hunting areas located there. This explains why the ranches that deforested a considerable part of their properties are among the most popular ranches used for *marisca* by the Angaité. This result of this scheme should be considered carefully. Deforestation implies the destruction of the habitats of animal and plant resources used by the Angaité. The reason why the Angaité hunt in ranches that deforest is only due to access loss in other areas. This strategy can only remain in use while the ranchers allow the Angaité workers and their visiting relatives to hunt in those areas or until those areas are also deforested.

Finally, in section 3.4, I presented the Angaité's additional comments on the process of territorial dispossession they experienced. Their words reflected their awareness of the magnitude of the changes happening in their region and the rest of the Chaco. The loss of access was mentioned again and again by the Angaité. This proved that "La Patria territory" is *de facto* not recognised by the state or the ranchers. I was attentive to how the Angaité expressed their cosmography in relation to comments on the changes considered here. On one hand, there is an alternative understanding of property/ownership/management of land, animals and forest. They refer to the forest historically used by the Angaité but currently located in private properties as "our forest" and add the clarification that "forests have no owners". They consider a deforested forest (a pasture) or a fenced property as owned by the ranchers. Meanwhile, wild animals and forests are considered to belong to the Angaité. The expression, *"the ranchers ended up owning everything"*, also reflects that the co-existence in the territory was not expected to imply enclosure and exclusion. On the other hand, there is a resemblance between the understanding of what "La Patria territory" is and the abovementioned definition of place. It is considered the result of the social interactions and practices of all the beings that are part of a broadly defined "social collective". This is why I argue that when Mr. Santiago says that "*When everything started to be deforested, it seems that the wild fruits season never reached again and the shortage started*" he is describing both a social and an ecological phenomenon. Therefore, the drastic process of deforestation is related to the new attitude and new kind of social relations. The shortage is considered the result of a break in basic sharing principles on behalf of the ranchers and other actors, including the Angaité. I am not claiming that the Angaité do not perceive the deforestation as an ecological phenomenon but that they highlight the Paraguayans' change of attitude towards the indigenous people as a cause. Life before this change is described as follows, "*the ranchers were not mean yet and we were tranquil*". For the Angaité, a tranquil everyday existence is still considered among the most important outcomes of their social relations. They know their rights but see the law as something that leads to confrontation between themselves and the ranchers.

The Angaité voices conceive "La Patria territory" as a place they inhabit along with animals, plants and other non-humans as well as with ranchers and "not so human" others. The change of attitude of the ranchers and their new ranching practices modified the social relations and changed "La Patria territory". The past period's balanced or good social relations between the Angaité, non-humans and Paraguayans are contrasted against the current period's conflictive and difficult relations. However, the present situation is not considered permanent or static. The lack of respect for their rights and the change of attitude towards them are attributed to a lack of understanding from the state and the Paraguayans in general. Thus, the Angaité are trying to re-establish the good relations by displaying who they are, where they live and what they eat. They are responding to this lack of understanding by actively trying to find a way to display their cosmographic practices, to communicate who they are and why their territory is important to them. Their historical connection to their territory through the *marisca*, following their parents and grandparents, are among the issues they consider important to communicate and understand.

If we consider, from a more "classical" Political Ecology perspective, the Angaité's current territorial struggle as a battle over natural resources and/or land we might conclude that impacts associated to environmental change are distributed unequally among actors. This is evidenced by the process of territorial dispossession experienced by the Angaité over the past 20 years. In this case the unequal actors are the state, the ranchers and the Angaité. By authorizing the deforestation without considering indigenous territorial rights and the existence of "La Patria territory", the state institutions failed to consider the externalities and direct environmental impacts on the Angaité of La Patria. This unequal distribution of environmental impacts reinforced the social and economic inequalities among these actors and reduced the possibilities of the Angaité to respond or resist the ranchers. This conclusion might be only partially correct, if the analysis stops here, as some crucial aspects of how the Angaité are understanding this struggle and their present moment are left aside.

This struggle is better understood if conceived as an ontological conflict, I define the latter as "*conflicts that ensue as different worlds or ontologies strive to sustain their own existence as they interact and mingle with each other.*"[103] I argue that by not recognising "La Patria territory", the Paraguayan state allowed and promoted resource use practices that were incompatible with certain Angaité livelihood practices. As the cosmographical practices contribute to the production of particular places/territories, by privileging practices associated with the "development" cosmography, the Paraguayan state disabled the production of "La Patria territory" and the understanding of the world associated to it.

---

[1] PNC, 2012
[2] Villagra 1998, cited in Andrew Leake, 1998
[3] Andrew Leake, 1998: 272
[4] Mario Barrios, 2008
[5] PNC, 2012: 24
[6] ILO, 2003

[7] Grupo Sunu, 2007
[8] Ibid.
[9] Ibid.
[10] Andrew Leake, 1998: 56
[11] Arturo Escobar, 1999
[12] Volker Von Bremen, 1987: 12
[13] Ibid., 12
[14] Ibid., 12
[15] Florencia C. Tola, Celeste Medrano and Lorena Cardin, 2013
[16] Bruno Latour, 2005
[17] Philippe Descola, 2005
[18] Alexandre Surrallés and Pedro García Hierro, 2005
[19] Ibid.
[20] Valentina Bonifacio and Rodrigo Villagra, 2016
[21] Rodrigo Villagra, 2010
[22] Kalish 2008 in Rodrigo Villagra, 2010: 202
[23] Kalish in Rodrigo Villagra, 2010
[24] Valentina Bonifacio and Rodrigo Villagra, 2016: 91
[25] Valentina Bonifacio, 2013a: 512
[26] Ibid.
[27] Valentina Bonifacio and Rodrigo Villagra, 2016
[28] Rodrigo Villagra, 2010: 145
[29] Valentina Bonifacio and Rodrigo Villagra, 2016
[30] Ibid.
[31] Kidd 1995 in Valentina Bonifacio, 2013a: 511
[32] Valentina Bonifacio, 2013a
[33] Valentina Bonifacio, 2013a; Rodrigo Villagra, 2010
[34] Rodrigo Villagra, 2010: 244
[35] Valentina Bonifacio and Rodrigo Villagra, 2016: 93
[36] Rodrigo Villagra, 2010; Stephen W. Kidd, 1995
[37] Rodrigo Villagra, 2010
[38] Rodrigo Villagra, 2014
[39] Ibid., 164
[40] Ibid.
[41] Florencia C. Tola, Celeste Medrano and Lorena Cardin, 2013: 58
[42] Ibid.
[43] Rodrigo Villagra, 2010: 209
[44] Ibid., 209
[45] Florencia C. Tola, Celeste Medrano and Lorena Cardin, 2013: 66
[46] Ibid., 76

[47] Alexandre Surrallés and Pedro García Hierro, 2005: 11
[48] Arturo Escobar, 1999
[49] Valentina Bonifacio and Rodrigo Villagra, 2016: 104
[50] Andrew Leake, 1998
[51] Ibid.
[52] Ibid., 73
[53] Ibid., 132
[54] Ibid., 133
[55] Ibid., 129
[56] Ibid.
[57] Ibid.
[58] Ibid., 250
[59] Ibid., 96
[60] Ibid., 271
[61] Ibid., 255
[62] Ibid., 267
[63] Ibid., 248
[64] Ibid., 248
[65] Ibid., 267
[66] Arantxa Guereña and Luis Rojas Villagra, 2016
[67] MAG, 2008
[68] Ibid.
[69] Marcos Glauser, 2009
[70] Arantxa Guereña and Luis Rojas Villagra, 2016
[71] Ibid.
[72] MAG, 2008
[73] Ibid.
[74] Luis Galeano, 2010; Arantxa Guereña and Luis Rojas Villagra, 2016
[75] Luis Galeano, 2010
[76] MAG, 2014
[77] Alberto Yanosky, 2017
[78] Guyra Paraguay, 2014
[79] M. C. Hansen et al., 2013
[80] Mario Blaser, 2003: 202
[81] Ibid., 203
[82] Arantxa Guereña and Luis Rojas Villagra, 2016
[83] Mónica Centrón, 2016
[84] Ibid.
[85] Alfredo Lampert, 2015
[86] Ibid., 30

[87] Ibid., 6
[88] Ibid., 22
[89] Mónica Centrón, 2016
[90] Ibid.
[91] Adolfo Aquino, 2016: 23
[92] Ibid., 24
[93] Ibid., 41
[94] Nicolas Bertram, 2016
[95] Ibid., 1
[96] Victoria Tauli-Corpuz, 2015: 7
[97] Joe Bryan, 2009
[98] DGEEC, 2004
[99] Carolina Tytelman, 2016
[100] Rodrigo Villagra, 2014: 33
[101] Valentina Bonifacio and Rodrigo Villagra, 2016
[102] Carolyna Tytelman, 2016: 245
[103] Mario Blaser, 2010: 11

4 Process of change of livelihood strategies (1995-2015)

4.1 Introduction

In this chapter I explore and describe how the livelihood strategies of the Angaité of La Patria changed in the past two decades. I do so in order to understand how they responded to changes in their access to, use of and management of natural resources inside and outside their colony.

I consider livelihood activities part of the practices that contribute to producing places/territories and particular understandings of the world. In this case, they contribute to producing "La Patria territory". This approach is influenced by Ingold's "dwelling perspective", according to which the perception of the environment and the knowledge of the world is obtained through a process of *enskilment* in the course of everyday practices.[1] Through this process of *enskilment*, or education of attention, *"cultural knowledge and bodily substances are seen to undergo continuous generation in the context of an ongoing engagement with the [human and non-human] beings that dwell therein."*[2] Following this argument, the loss of access to certain places and the abandonment of certain practices (or the adoption of new ones) must impact how the Angaité transmit their cultural knowledge and perceive their territory. In this chapter I show that the Angaité responded to the enclosure of "La Patria territory", caused by the deforestation and access restrictions, by: adapting their livelihood strategies; decreasing the importance of some activities/practices (hunting, fishing, gathering); and increasingly depending on state social programs as well as on livelihood activities/practices conducted inside (land leases and timber extraction) and outside (wage labour in cattle ranches) the colony.

I use the following definition of the livelihood strategies concept, a combination of activities that people choose to undertake in order to achieve their livelihood goals. Livelihood goals refer to making a living, trying to meet economic necessities, coping with uncertainties and responding to new opportunities.[3]

Regarding the livelihood activities of hunter-gatherer societies, Ingold suggests the notion of *procurement*, which involves management, care and knowledge. He argues that livelihood practices in such societies should not be reduced to behavioural responses or external environmental stimuli, as implied in the notion of foraging. Nor should they be regarded, as implied in the notion of production, as planned interventions in "nature" launched by "society".[4] In his definition of procurement, Ingold also emphasizes that non-human agents are involved in these practices: "*the activities we conventionally call hunting and gathering are forms of skilled, attentive 'coping' in the world, intentionally carried out by persons in an environment replete with other agentive powers of one kind and another.*"[5] This resonates with a central element of Von Bremen's argument about hunters and gatherers of the Chaco region. He stated that for the indigenous people of the Chaco, magical (ritual) and practical (hunting) techniques have a mythological fundament: "*all natural and cultural phenomenon have a spiritual aspect, they possess their own individual personality… and even their own will.*"[6] A similar argument has been then further developed by other scholars. [7] Von Bremen also suggested that the indigenous people of the Chaco, like other hunter-gatherers, think that the resources needed to survive are abundant and pre-existent; therefore, resources need to be appropriated rather than "produced".[8] In his critique of Von Bremen's argument, Gordillo expressed his concern for the "*distinct ontological status*" attributed to the indigenous people from the Chaco, "*the status of those who are not productive, those who recollect, who are not living 'of nature' but 'in nature'…*"[9] I explored the process of change of livelihood activities keeping this discussion in mind. I come back to it in the conclusion of this chapter.

The findings presented here are based on the participatory mapping, observations and interviews conducted in the 18 different villages of La Patria. However, special attention is paid to the particular situation of four villages: *Karoa'í*, Tataré, *San Fernández* and *Tres Quebrachos.* Again, I compare my findings to the situation Leake described in 1995. A second anthropological research conducted there, by Rodrigo Villagra, is also a rich source of information. These two sources are enriched with information from the Indigenous Population Censuses conducted in 2002 and 2012.

4.2 Resource areas used by the Angaité inside La Patria colony

Despite the changes considered here, and certainly in different degrees of intensity, in 2015 the Angaité were still using all the resource areas used in 1995. The resource areas used include (see Table 6) natural grasslands, man-made pastures, water reservoirs (*tajamares), viñal* forests (*Propopis ruscifoglia)*, algarrobo forests *(Prosopis),* flooded lowlands *(bañados),* lagoons, marshes, palm savannahs, gallery forests, streams, *quebracho* forests (*Schinopsis balansae),* thorn scrubs, palo santo forests (*Bulnesia sarmientoi*), *mistol* forests (*Ziziphus mistol*), and *labón* forests (*Tabebuia nodosa*). Some of these vegetal formations were described in section 2.2.1

Table 6. Types of resource areas used by villagers inside La Patria colony for their livelihood activities

Activity/ Resource	Natural grassland	Man-made	Water reservoirs	Viñal forest	Algarrobo forest	Flooded lowland	Lagoon	Marsh	Palm savannah	Gallery forest	Stream	Quebracho forest	Thorn scrub	Palo santo forest	Mistol forest	Labón forest
Settlement									x			x		x		
Horticulture	x									x						
Cattle grazing		x			x	x	x		x							
Wild fruits gathering	x				x	x	x		x	x	x				x	
Honey collecting					x	x			x	x		x	x			x
Fishing			x				x	x			x					
Hunting	x		x	x	x	x	x	x	x	x	x	x	x		x	x

These resource areas were used in 2015 for settlement, horticulture, cattle grazing (which includes their own cattle, and the cattle that belongs to non-indigenous persons leasing land), honey collecting, fishing and hunting. As Leake observed in 1995, the resource areas used for cattle grazing overlap with the most important resource areas used by the villagers for hunting and gathering. Considering that the cattle population increased substantially due to the land leases, the number of conflicts generated by the land leases in the past years is not surprising.

The Angaité's use of these resource areas involves the transformation of the "natural" condition of these ecosystems. The continuous and repetitive uses create modified ecosystems that benefit certain species and not others. However, there are few practices that can be considered part of a conscious and planned management of the natural resources. The annual use of fire is among the few exceptions. The Angaité say that they burn the grasslands in order to attract wild animals that feed on the renewed grass. The cattle also eat this renewed grass. People also argue that if they do not burn the grassland it turns into *maciega.* If other vegetable species invade the grasslands, they become *maciega (oñemasegá)* that impedes the movement of people and animals. In many occasions, the Angaité related the *maciega* to the abundance of "mean animals" (snakes). The annual burn of grasslands is done from July to October/November but depends on the rains and wind directions.

The use of the abovementioned resource areas is illustrated by the results of the participatory mapping conducted in *Tatarē* village, composed basically by the leader and his extended family. The same mapping exercise was conducted in every one of the 18 villages of La Patria. In the case of *Tatarē* village, land and resource use (inside the colony) did not change considerably in the 20-year period considered here. It is still centred on the core area immediately surrounding the site of residence, where the horticultural garden and beehives are located. *Tatarē* village core area is situated on an elevated grassland, close to a stream and not far from different kinds of forests. This means that the settlement covers all the fundamental ecological characteristics of a traditional Angaité settlement.

The goat and sheep barn are beside the leader's house. Not far from this core area lies the village paddock that is used in common by all households. *Tatarē* currently has no cattle, thus, the paddock is used to keep horses and donkeys. However, during the year I spent in La Patria this and most village paddocks were empty. The cattle, horses and donkeys grazed free in the colony lands. Next to the paddock are two big water reservoirs (*tajamares).* Water from these reservoirs is used to bathe, drink and fish. To bath, people do not go into the reservoir. An abandoned agricultural plot is between the houses and the paddock. Wild fruit and plants are gathered within a broad area surrounding the village. Palm hearts were gathered in the areas of palm savannah and *algarrobo* was gathered in the *algarrobo* forest next to the streams. In this case, the areas used are almost the same as those used 20 years ago; however, the frequency and intensity has varied.

Figure 11 also shows areas used for hunting, fishing and gathering. During the mapping exercise, the villagers found it difficult to indicate where they hunted. Perhaps because the areas used for hunting include almost every kind of resource area. However, specific spots used for fishing were easily identified. The resource areas used by *Tatarē* were also used by other villages. In fact, only the village core area was not used by villagers from other villages. Villagers from neighbouring villages hunted, gathered and fished in the area that according to the *Tatarē* leader correspond to *Tatarē* "village territory". Timber extraction and land leases were the exception, only villagers from *Tatarē* extracted wood or were involved in land lease deals in this territory. As it will be shown, it was not always clear what areas correspond to each village. Therefore, the intensification of timber extraction and land leases also created potential conflicts.

Figure 11. Sketch map of *Tatarē* village

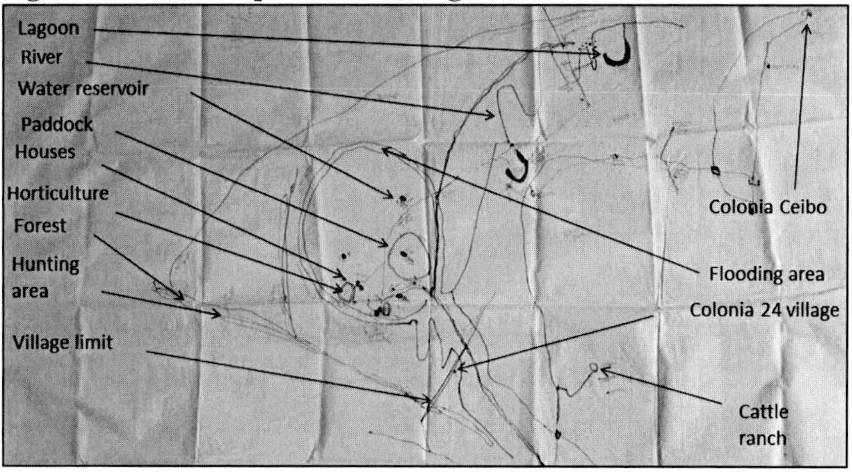

4.3 Livelihood strategies of the Angaité of La Patria

In order to overlook the livelihood activities in which the Angaité of La Patria were involved in 2015, I surveyed tools and resources in 19 households from four different villages. The idea behind the survey was to extract an initial overview of the extent to which households were geared towards engaging in different activities by calculating the extent to which they owned or had access to these tools and resources. In order to estimate if leaders are accumulating tools and resources as well as to compare the situation in four different villages, I included one leader's household per village and a total of 15 randomly selected households in the survey.

Table 7. Household ownership of tools and resources required for livelihood activities

Village	San Fernández				Karoa'i						Tataré					Tres Quebrachos				Total
Household ID	1	2	3	**4**	5	6	7	8	9	10	**11**	12	13	14	15	**16**	17	18	19	
Ranch work	N	N	Y	Y	Y	N	Y	N	Y	Y	N	N	N	Y	N	N	N	Y	Y	9
Horse			1	3							2									6
Donkey	4	2		3	2			1	4	2	7		10	1	1					37
Cattle				29	14															43
Sheep/goat				5							80	2	2							89
Pig																				
Chicken	6	5			30		5	1	4	5	20	5	4	4			8	8	10	115
Beehives		4		1							3		1							9
Apicultural equip.	N	Y	N	Y	N	N	N	N	N	N	Y	N	N	N	N	N	N	N	N	3
Garden	Y	Y	Y	N	Y	Y	Y	N	Y	N	Y	Y	Y	Y	Y	Y	Y	Y	Y	16
Garden fences	Y	Y	Y	N	Y	Y	Y	N	Y	N	N	Y	Y	Y	Y	Y	Y	Y	Y	15
Axe	3	2	2	3	1	1	2	1	1	1	3	2	2	1	1	2	2	1	1	32
Machete	4	2	1	4	1	1	3	1	1	1	3	1	2	1	1	1	1			29
Spade/hoe	5	3	2	3	9	1	2	2	2	2	10	3	3			5	4	3	5	64
Hunting dog	5	3	1	2				1	1		2	1	1			2	3	1	3	26
Rifle				1					1		1		1							4
Revolver				1							1									2
Fishing hook	Y	Y	Y	Y	Y	Y	Y	Y	Y	Y	Y	N	Y	Y	Y	Y	Y	Y	Y	18
Fishing trap	2	2	1	1					2		2		2	2	2	1	1	1	1	20
Spear	2	2	2	2	2		1	1	2	1	3	1	2	2	2	3	2	2	2	34
Motorbike				1	1		1			1							1			5
Chainsaw	1	1		1	1	1	1	1	1							1				9

Highlighted households (4, 6, 11 and 16) correspond to leaders of each village.

Table 7 shows the results of this survey. In the table, possessions have been grouped according to the livelihood activity they are associated with. However, there is some overlap between activities that involve the same tools. Axe and machetes are tools used for horticulture but also for hunting. Spears are used as hunting and fishing tools. Horses and donkeys are mainly a mobility resource, the latter are also used for hunting sometimes. I found that 9 out of 19 households were involved in wage labour on cattle ranches. Cattle ownership was concentrated in two households, both belonged to village leaders. However, they clarified that the cattle were of communal ownership. Sheep and goats were mainly concentrated in one leader households and were not of communal ownership. As in 1995, no household owned pigs, but the average of chickens owned per household multiplied by three. In this case, the highest number also corresponded to leader's households. Only 4 out of 19 households were involved in beekeeping and only three owned some apiculture equipment. This clearly differs with the situation in 1995 when a considerable number of households were involved in apiculture. The majority of the households have a home garden and the tools used for that activity. As in 1995 in *San Fernández* the leader's household didn't have a home garden. Interestingly the highest number of axes, machetes, spades and hoes are also owned by the leader's in all the selected villages.

There was an apparent difference in how these households got involved in hunting. Hunting dogs are mainly concentrated in two villages. However, the villages with lower number of dogs are the ones where some households also owned rifles or revolvers. It has to be clarified that in *San Fernández* the leader was mainly using these guns to protect the cattle and for personal security, not for hunting. Most of the households were equipped to get involved in fishing. The number of chainsaws and motorbikes per household and village differed, *Karoa'í* and *San Fernández* had the highest amount. The number of chainsaws in *Karoa'í* evidences that these households depended more on timber extraction than the other villages.

In 1995, Leake found that the majority of the surveyed households were equipped to engage in the 7 identified subsistence activities: wage labour, livestock raising, beekeeping, horticulture, hunting, gathering and fishing. That year, the extent to which households engaged in livestock raising and wage labour on ranches differed.[10] I found two new activities that are combined with the ones identified by Leake, these are timber extraction and land leases. I found that the majority of the households were equipped to engage in most of these activities, excluding beekeeping and land leasing.

Let me now describe in detail each one of these livelihood activities/practices.

4.3.1 Marisca

Marisca is a term used in Paraguay and the whole Gran Chaco to refer to hunting, fishing and gathering activities. These activities are highly interrelated and can hardly be separated.[i] Despite the deforestation and restricted access to ranches that overlap with "La Patria territory", the importance of *marisca* in 2015 was clearly expressed by one leader of *Carpincho* village. The coordinator of a UNDP executed project asked, "*What do you need here?*" and the leader answered, "*we depend on the marisca, but our neighbours are not allowing us to enter (their properties) to kill not even a long-nose armadillo. When the weather allows it, we collect and sell honey.*"[ii] Hunting, fishing and gathering are highly dependent on weather conditions. These activities require more effort during the flood season. Thus, people prefer the dry season and describe it as a period of abundance.

4.3.1.1 Hunting

The last Indigenous Population Census, conducted in 2012, evidenced that 73% of the households of La Patria stated that at least one of its members was regularly involved in hunting for subsistence.[11] Out of these, 89.7% said that the purpose of hunting was to consume the meat. They specified the most popular preys: anteater (*Myrmecophaga tridactyla),* long-nose armadillo (*Dasypus novemcinctus),* peccary (*Tayassu tajacu, Tayassu pecari),* iguana (*Tupinambis teguixin),* stork (*Euxemura maguari),* capybara (*hydrochaeris hydrochaeris),* caiman (*Cayman spp.),* grey brocket deer *(Mazama guazubira), coati (Nasua nasua),* turtle, and *ñandu (Rhea Americana),* among others birds.[12] In 1995, Leake also mentioned the long-nose armadillo, both types of peccary, the iguana and the deer as the most frequently hunted species.

Leake differentiated five types of hunts: "opportunity hunts", "crepuscular hunts", "day hunts", "expedition hunts" and "hunting while working on ranches". The first one is carried out when wild animals happen to be encountered during the course of travelling. This type, probably the simplest one, currently still takes place. While traveling with the Angaité to or inside the colony I observed that they were always ready to catch or shoot any encountered animal. I witnessed how a rifle was used to hunt a tapir (*Tapirus terrestris spegazzinii)* and how a revolver was used to hunt a long-nose

[i] *Not only hunting, fishing and gathering are very interrelated, when someone mentions that he/she was working (changa) in a cattle ranch, most of the time he/she also means that he/she has been hunting and collecting honey while working there.*

[ii] *"…nosotros somos mariscadores, pero acá nuestros vecinos no nos dejan entrar a cazar ni un tatú. Acá a la miel nos dedicamos cuando da el clima." Mr. Daniel, May 2015*

armadillo. When walking for long distances, coming back or going to a ranch, they were usually armed.

The "crepuscular hunt" is carried out in most of the villages, yet not all household have a hunter (*mariscador)* among them. In *Karoa'í* village, the leader's daughter´s husband used to hunt with a rifle in the early morning or evening, for 2 to 3 hours. He made a round trip to the north of the village, following the river towards *Karova Guasú* village, and returned across the forest to the West of his village. When he caught a good prey, though he was not successful every day, he shared it among 3-4 households. When planning to enter areas with thorny shrubs like the *viñales (Prosopis ruscifolia)*, he sometimes took his well-trained donkey. His most common prey were the *charatas (Ortalis canicollis).* In other villages, men without guns used hunting dogs. The general perception is that the number of animals inside the colony decreased over the past years. Among all the villages, five are located in favourable places that have easy access to the less populated areas of the colony. These are: *Laguna Hũ*, *Monte Kué*, *Carpincho*, *Tatarẽ* and *Laguna Teja.*

I observed how some men dedicated their entire day to foraging activities, what constitutes a "day hunt". This type is still practiced inside the colony but has decreased in surrounding properties over the past years. The rapid deforestation and increasing access denial of the surrounding properties contrasts the relative freedom people had in 1995. Despite the fact that the colony was already surrounded by *Estancias,* in their words, at that time everything was considered "open land" (*campo abierto).* Though not so frequently, I still observed people traveling by horse or by motorbike to hunt along the road towards *Colonia Ceibo* or towards *Puerto Pinasco.* I also witnessed, in *Tres Quebrachos* village, how four men, most of them relatives, spent more than 5 hours trying to catch a long-nose armadillo. The time and energy invested reflects the importance they attribute to this kind of meat.

The "expedition hunts" share the same context of deforestation and loss of access to the *estancias.* This fourth type mentioned by Leake, involves groups of people over the course of several days, normally establishing a camp in the forest, is also less frequent. People mentioned *"this is how we used to live in the past, staying away for 15 days, sleeping in temporary camps while collecting honey and other things to sell or exchange for supplies. But happily now we get the emergency kit and we can stay at home and calm down for a few days."*[iii]

The fifth and last type of hunt refers to people that engage in hunting while undertaking wage labour or visiting a relative living on a ranch. Leake also mentioned that "*this practice often allows the Indians to circumvent the hunting prohibitions imposed by landowners.*"[13] This is the most controversial change related to hunting in the last years. Due to the decreased number of wild animals inside the colony, there is a tendency to hunt where people used to live in the past. These places are considered their *tekohague,* "their past community" or "their place to be."[iv] As one leader from *Karoa'í* village mentioned, *"it was there where they lived, they know how to enter to the forest there, where to look for the animals, animals from the forest and honey."*[v] However, today most of these *tekohague* are located within private properties. This return to places inhabited in the past also coincides with the increasing restrictions on ranch worker's permission to host guests or hunt.

Leake mentions the following hunting tools: dogs, different types of spears, rifles, revolvers and clubs or heavy sticks. I observed more sophisticated shotguns and a decreased diversity of spears. Basically, the only common spear is the one used for catching eels (*Synbranchus marmratus)* and armadillos (see picture 4). The use of clubs or heavy stick is not frequent. However, I witnessed men from *Carpincho* village use torches and flashlights on a night hunt for storks (*Euxenura maguari)*, locally called *tuyuyú,* at the lagoons.

Hunting becomes a very difficult activity during a flood. The reduced number of wild animals inside the colony contrasts with their abundance in some higher elevated places located inside the private *estancias.* Non-indigenous Paraguayan workers commented on the abundance of the *tajykatí,* the Guaraní name for the peccary (*Tayassu tajacu),* during the long flood in 2014. Sometimes, thanks to friendship links or kin relations among the

[iii] *"Así era como se vivía yma, ausentándose durante quince días, durmiendo en ranchos temporales acumulando miel y otras cosas para vender a la vuelta o cambiar por provistas, pero por suerte ahora ya osema la emergencia y podemos quedar mas tranquilos sin irnos lejos por muchos días". Mr. Daniel, August 2015. A detailed description of the emergency kit is available in the state related income section of this chapter.*

[iv] *The word used by them is tekohague. Tekoha is commonly used today as synonym to community. However, teko is also translated as "way of being, law, culture", and tekoha as "the place where one can be" Bartomeu Meliá and Adriana Almada, 2011. Tekoha is therefore very much related to the concept of territory.*

[v] *"Ko'ápe ningo sapy'ánte opa, ohopa ka'aguýpe, hekohague voi. Imandu'a hekohaguépe. Upépe oiko ha'e. Upépe oikuaaa moõpa oike pe ka'aguýpe, mo'õpa oheka la animal, vícho ka'aguy umíva. Upépe voi oiko, ha upépe jeýnte oho va'erã ogueru haguã la vícho ka'aguy térã eíra." Mr. Santiago, August 2015*

Paraguayan workers and the Angaité from the colony, easily caught preys were shared. During the 2014 floods, given the lack of government support, people consulted the elders on where to find caimans. Special hunting expeditions were organized to a small dam, located close to Estancia 40, where they would catch more than 10 caimans in a few hours.

There are empirical hints towards what seems to be a very well hidden local ontology. One of the leaders of *Karoa'í* village mentioned that he is not hunting anymore because he encountered the *Jary Ruvicha* in the forest. He described this "being" as a huge deer and explained that for a hunter this encounter signals that he already hunted too much in his life. When interviewees describe the dangers involved in hunting during a flood they avoid mentioning snakes, referring to them only as *mean animals*[vi] (*vicho vaí)*, but recurrently mention the shaman that "talks to fishes".

4.3.1.2 Fishing

An old woman made the following statement to me during my first visit to *Carpincho* village, *"We eat two times a day, sometimes only once a day. When there is no food, fish is what keeps us alive."*[vii] Fishing is probably one of the most reliable sources of food in La Patria. About 87% of the households stated that fishing was an important livelihood practice in 2012.[14] The most frequently captured species in 2015 included *tare'y* (*Hoplias malabricus*), a type of eel locally called *mbusú (Synbranchus marmratus), mandi'í (Pimelodus spp.), piraña (pygocentrus pyraya) sábalo (Prochilodus spp.)* and *cascudo (Plecostomus punctatus)*. The *pacú* (*Calosum brachypomus),* mentioned by Leake as a common species, was not mentioned by my respondents nor did I see one.

The techniques used for fishing differ according to the water level in the rivers, lagoons and water reservoirs (*tajamares)*. People tend to use a hook and line with high water level. When water level decreases, people fish in large groups, using a spear *(fija)* and conical traps (see picture 3). This trap, locally called *palo cavajú,* is constructed out of thin strips of wood. When using the conical traps, men and women form a line across the pools along the river, locally called *pozones* and walk towards a shallower part of it. In the shallower part, kids slap the water with branches to scare the fish. Each individual keeps the trapped fish in a sackcloth bag secured to his waist. Not every group

[vi] *"Ha la vícho vai heta oĩ ko zona. Altura'i oho lo mitã ha upépe oĩmba umi vícho, ojo'a"*

[vii] *"Comemos dos veces por día, a veces una. Cuando nos falta comida, la pesca es lo que más nos salva" (Mrs. Lisa, June 2015)*

fishing expedition is successful. The changing weather is making it even more difficult to decide when and where to fish. During the flood season fishing is both more complicated and more necessary as the possibilities of hunting, going to ranches to work or even buying food are reduced. I observed a large group of men and women, from almost every household of *Karoa'í* village, try a new fishing technique: they used a nylon fishing net that traversed the whole river. After hours, they decided to give up without catching any fish. Two days later, the water level was so low that the fish became weak. People could easily catch them using a machete and their hands. According to the Angaité, they are fishing less now than in the past years. In *Carpincho* village, people related this fact with the death of a famous shaman called Chamorro, which used to "talk to the fishes" at the lagoon. After hours of failing to catch anything, he used to tell the people to start again. According to many Angaité, thanks to his interventions, people were catching many fishes, caimans *(cayman spp.)* and capybara in a few minutes.

As riverine villages are more dependent on fishing, they have more fishing related knowledge, know more about fish behaviour and best fishing techniques for specific contexts. A man from *Karoa'í* village invited families from *San Fernández* village to join a group fishing expedition to the river in October. These families refused because they were expecting to receive payment that day in their village from the *Tekoporã*[viii] program. This indicates that the risk of missing the payment is higher than the potential benefits of a good fishing expedition.

People are also fishing outside La Patria colony in the small water reservoirs formed along the roads, in rivers and water reservoirs inside the *estancias.* However, as said, access to *estancias* has become increasingly prohibited in the last years. Women and children from *San Fernández* and *Tres Quebrachos* organised fishing expeditions. They walked for more than 15 kilometres along the dirt road to *Colonia Ceibo* to fish under the bridges. A strange phenomenon occurred in the past years. Thousands of fishes were found dead in the river. People were unable to provide an explanation. Some leaders however related it to the use of chemicals for weeding the cattle ranches.[ix] Thus, local leaders are aware of the problems they will face if

[viii] Governmental Conditioned Monetary Transfer program. See more on state related income section.

[ix] In July 2017 the leaders communicated again that due to the use of "round-up" in nearby estancias a massive death of fishes occurred again.

intensely sprayed (with fertilizers, herbicides) soya is produced in the nearby *estancias.*

A fishing calendar was constructed based on observations and interviews. The floods seem to have changed the scenario. Fishing was conducted year-round, not only from June to November, as Leake observed (1998). However, all group fishing was conducted from May and October. Despite the fact that households are currently less dependent on fish than in 1995, it is still a very important source of food.

4.3.1.3 Gathering

Wild honey is an important source of calories in the local diet. It is also an important source of income in the summer because wage labour in cattle ranches decreases due to rain and extreme heat. As in 1995, there were many species of honey-making insects, among them, the honeybee (*apis mellifica),* the *tumbypará (Meliipona quadrifasiata*), the *jate'í* (*Tetragonisca angustula)* and the *tapesuá (Trigona bipunctata).* The wild honey produced by the honeybee is the most important in quantity. It is collected inside and outside of La Patria colony. Gathering wild honey, it normally involves cutting down the threes, has increased the difficulty of finding good quantity and quality honey inside the colony. At the same time, honey collecting expeditions which allowed them to travel and stay for a couple of days inside the surrounding properties are less common today due to the access restrictions. However, people from different villages are traveling considerable distances by horse, donkey, motorbikes, bicycles or even on foot in order to get to specific places to collect wild honey. When selling the honey, the name of the *estancia* where it was collected is usually indicated on the label to specify its origin.

Despite the repeated attempts of many development projects to promote apiculture in La Patria over the past 20 years, the number of beehives is currently considerably low in the colony. Only 8 individual producers, from six different villages were identified.[x] Some of them received beehives from the Anglican projects more than a decade ago. Others were trained in the context of *Prodechaco* or more recently by *APICSA.*[xi] Leake attributes this low rate of adoption to the fact that beehives restrict the individual freedom to

[x] *Tataré, Colonia 24, Puente Kaigue, Carpincho, Karoa Guasú, La Leona*

[xi] *Prodechaco was a huge development program supported by the European Union and the Paraguayan government. APICSA (Apicultura del Chaco Sociedad Anónima) is a Mennonite enterprise that promotes apiculture among indigenous communities.*

engage in other subsistence activities since it requires the full presence of the owner. This is true in the Chaco context. Constant surveillance is the only way to prevent wild animals from destroying the beehives. During my fieldwork, new attempts were made by Grupo Sunu, they supplied financial support and training for the creation of a local sawmill where beehives would be constructed, as well as the UNDP related project. According to the apiculture technicians of the latter project, the floods, combined with the fact that human settlements are located in the highest elevated areas, reduces the total number of possible beehives in the colony to only 200 units.

The *macateros* (traveling merchants) are the main buyers of honey in La Patria colony. Although the purity and quality of this honey is high, the price paid by the *macateros* is probably the lowest in Paraguay. They use their empty trucks to transport the honey to the main national markets. There is no fixed price. Each *macatero* establishes his own price. During the 2015-2016 season, prices ranged from 15.000 G. (2.40 €[xii]) to 17.000 G. (2.76 €) per litre. The average price paid for honey of this quality in Asunción or Concepción at the same time was of 30.000 G. (4.80 €). Payments in kind (rice, pasta, oil, sugar, etc.) are common though the Angaité disfavour them. Besides the *macateros,* people from *APICSA*, Grupo Sunu and an Anglican technician also occasionally buy honey, paying roughly 23.000 G. (3.74 €) per litre. In *Tres Quebrachos* village one household collected honey from December to February, inside *estancia Santa Rosa.* Most of the time the working team consisted in two adults and one 12 years old child that helped make smoke to pacify the bees. They estimate that they sold 30 litres of honey in this time span, which seems credible or even too little. As Leake mentions, the Angaité are not keen on disclosing the amounts harvested because this would disclose how much money they got.[15]

People from La Patria also gather wild fruits and plants for different purposes. Palm trees (*Copernicia alba)* for example are used as construction material for houses. In fact, 60% of the walls and 6.7% of the houses' roofs were made of palm trunks in 2012.[16] The palm heart is consumed and the palm leaves are used as materials in handicraft. The importance of this plant was already observed by Leake. Apparently, its status didn't change. Another historically important tree is the algarrobo (*Prosopis spp.)* which is a favoured fire wood. The quebracho (*Schinopsis balansae)* is sold or used for fences and construction. Not only palm leaves are used for handcraft. The *karaguata'í*

[xii] *Exchange rate of 01.09.2016 (1€ = 6,143 G)*

(Eryngium elegans) is used to make a very resisting fibre. This plant is gathered carefully, not uprooting the whole plant, allowing the remains to regenerate. Besides the aforementioned plants, handcraft is also made of wool, cotton, feathers, wood and mud.

As the Indigenous Population Census from 2012 states, 70% of the households of La Patria consume at least the following six wild fruits: palm hearts, *algarrobo* bean pods, prickly pear, coconut, *ñangapiry* and wild beans. Almost all the villages stated that they consume a similar list of wild fruits and plants. All besides *La Paciencia* village that only mentioned *algarrobo* bean pods.[17] This was almost the same list in 2002, when they also included *jata'i.*[18] Leake mentions 15 different types of fruit used for food in 1995, but warns *that "many of the interviewed people seemed shy and reluctant to disclose the types of fruit that they ate."*[19] The attitude of the respondents may be explained by the fact that Leake was associated with de Anglicans, who historically related the traditional food with de devil. In fact, I experienced a similar reaction in 2015 when an interviewee answered that in his village (*Urunde'y*) nobody consumed wild fruits because *"that is Satan's food".*[xiii] Trying to show his personal position regarding this demonization of wild fruits, an indigenous teacher from *Paraíso* village publicly stated *"I, despite being a teacher, eat the traditional fruits."*[xiv]

In 2015 wild plants or fruits appeared to be consumed mostly when no other food was available and the importance of wild species for the local diet decreased considerably. Leake mentions that during the *algarrobo* season (between December and February) women typically spent 4 to 5 hours a day gathering *algarrobo* pods from the forest floor. During this time of the year, I only saw women collecting *algarrobo* pods on 3 occasions. In different instances, people mentioned that the *algarrobo* pods do not ripen properly any more as a result of the distorted rainfall pattern. Others relate this phenomenon with the increased number of cattle ranches in the region. This last statement could be logically related to deforestation, but may hold a deeper meaning, it could be related to the impossibility of continuing a traditional way of life and to a change in the local social relations.

Women tend to underline the importance of *algarrobo* pods and palm hearts during some periods of the year, such as the flood season. An old woman mentioned to a technician of the World Food Organization, *"We eat*

[xiii] *Mr. Agustín, October 2015*

[xiv] *"Che, siendo profesor ha'u la fruta ymaguare" Mr. Andrés, August 2015, my emphasis*

palm heart and a lot of algarrobo pods and fish, that is what keep us alive"[xv]. He was conducting a "participatory rural appraisal" in *Carpincho* village. The importance of the palm heart for the colony's adaptive capacity to severe weather events, like the flood that affected La Patria in 2014, can be understood by the following account of a woman from *San Fernández* village:

> *"Here we made a small dam to stop the water, there are many mean animals (snakes). Now there is less water. We suffered a lot going to get some palm hearts. There was nothing to eat, no rice and that kind of food. We suffered. There were sick children and hungry women. We were all weak. (…) The whites didn't care. We survived the flood thanks to the palm hearts and the ysypo (liana). Very far we went, every morning. We did not rest even one day. Because there was nothing, the garden was destroyed by the flood. All the crops were destroyed. That is how we live here, at the colony. This is how we live. We suffer during the flood. We are not even scared any more of those animals (snakes). We step right next to them when we go to the palm forest full of leaves. But we are many, all women, I am not alone. Like in a demonstration we walk along the flooded road to the palm forest. Then we cook it, as a manioc. We only use the inside upper part of the palm, as traditionally we used to do. Culture, as some people say."*[xvi]

The account speaks for itself, but a couple of things should be noticed. Women are responsible for gathering wild fruits and plants, even during floods. She mentions bad animals to avoid mentioning the snakes by their own names. The indifferent attitude of the whites refers to the lack of emergency aid from the government. The concept of culture is used locally to refer to as a past "traditional" way of life. Despite the apparent decrease of importance attributed to wild fruits and plants in the current diet, these

[xv] *"Comemos karanda'y ru´a y mucho algarrobo y pescado eso es lo que nos salva".*

[xvi] *"Ápe rorrepresa, heta la vícho vai... Ko'ãga ningo sa'ive ou (la y), heta rosufri roho karanda'y ra'y roipe'a. Ndaipóri jakaru haguã, arroz umíva ndaipóri... Ore rosufri. Oĩ hasýva mitã, kuñakarai ivare'a. Ha ikangypa. (…) Noma'ẽi la karaikuéra (…) Ore karanda'y'ru'ã ha ysypo, umíva la ombojehasa la y heta. Ha mombyry, amo Buena Vista ipýri, laguna vera upépe roho roipe'a. Ha mombyry ningo upépe... Santa Rosa ruta roho, roguahẽ meve... Mba'e rojapóta, romongaru haguã la ore família kuéra roho rogueru... cada ko'ẽ... ore un día ndorotongueái. Porque ndororekói mba'eve, ndaipóri kokue pegua, la y heta, ojukapa...péicha ore rovivi ko colonia. …. Péicha ore roiko ko'ápe. Heta rohasa asy sapy'ánte péicha la y heta. Ore ndorokyhyje véima umi vícho, asulado rohasa la y heta, ore ningo iky'ahápe, karanda'ytýpe roho, ore retakue, kuñakarai, ndaha'éi che año. Ore ko'a pegua roho, romanifesta ruta ári roho karanda'yhápe roipe'a haguã. oñekosina, jetýicha. Oĩ ningo umi karanda'y apytere, upéva ore roipe'a, oĩ hu'ã, yvate,... upéicha guare ningo ore indígena roha'ã rokaru haguã, "cultura" he'íva ñe'ẽ.."*

have a critical importance for survival during floods or other situations of food scarcity.

4.3.2 Horticulture

4.3.2.1 Home gardens

The participatory mapping conducted in all 18 villages of La Patria colony identified 364 households and a total of 185 home gardens (*chacras*). This means, statistically, only 50,8% of the households have gardens. In fact, some households have more than one garden so the actual number of households without garden is even higher. People that belong to different household sometimes work together "as a society" (*en sociedad*). A household that apparently has no garden can join one of these joint ventures in order to maximise effort and make the most of the far from abundant horticultural land. Only one quarter (24,3%) of the gardens are fenced with wire or dried shrubs to protect them from the cattle. In the rest, one or more members of each household are in charge of looking after and protecting the garden most of the time.

The most important tools for horticulture are the hoe, spade, *foise*, axe and machete. Besides the tractor that is ploughing some gardens every year, most horticultural activities are undertaken manually. Most gardens are located in grasslands (*espartillares*). Only a small number of gardens are new slash-and-burned areas (*rosado*) in gallery forest. Most of them are located in recently created settlements, like *Karova Pyahu*, to the West of *Karova Guasú* village. In *Paraíso* village, people opening new plots stated that *rosado* in the gallery forest are an adaptation option to cope with the extreme weather conditions. In their words, in the high forest the seeds and the plants "*don't feel anything*".[xvii] This is probably due to the fact that high forests are located in elevated areas and the vegetation protects the garden from the heavy rain, sun and wind.

Due to the high population number in La Patria and the regular floods, there is a current shortage of horticultural land. The need for more horticultural land is one of the common arguments used by people creating new settlements in remote grasslands. This was different 20 years ago, when the availability of land allowed people to fallow their gardens after two or three years of use.[20] Considering that the average size of the gardens is 767,3

[xvii] *"noñandúi mba'eve" Mr. Leonardo, September 2015*

m², there is an estimated total of only 14,1 hectares of cultivated land in the colony. However, not all the villages cultivate in the same amount. This is shown in Figure 12, where the extension of land cultivated per person is estimated by dividing the average cultivated extension of each village by the total population according to the 2012 Indigenous Population Census.

Figure 12. Area cultivated per person in each village (ha) 2015

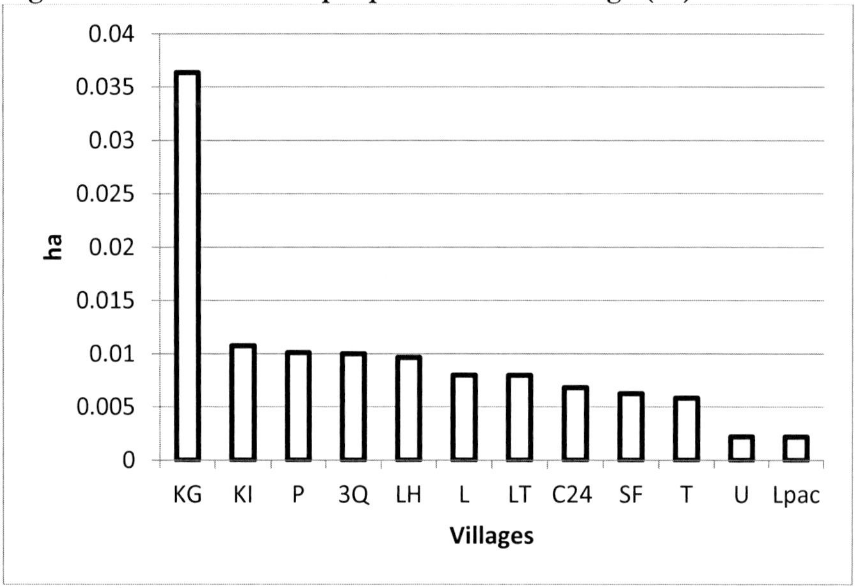

Figure 12 shows a difference of the total extension cultivated per person in each village.[xviii] The households from *Karova Guasú* are clearly dedicating more effort to horticulture than the households from the other villages. A second cluster is composed by *Karoa'í, Paraíso, Tres Quebrachos* and *Laguna Hū* villages which are quite even. The relatively low extension cultivated in *Urunde'y* village can be explained by the fact that most of the household depend on wage labour in cattle ranches. In *La Paciencia* village, the salty soil is known to be inappropriate for horticulture. Residents claimed to be tired of being accused of laziness by state and NGOs technicians that ignore their soil properties.

[xviii] *Some villages are not included because I lack enough information or because there were created after 2012, when the Indigenous Population Census was conducted.*

During the participatory mapping I observed all 185 home gardens. A list of the most common crops was elaborated based on this observation. I decided not to ask the people what type of crops they planted because they include what they "normally" plant along with what they "wish" to plant. Figure 13 shows the types of garden crops from September to November 2015. Sweet potato *(Ipomeas pp.)* and manioc *(Manihot spp.)* were the most widely cultivated staple. They were followed by watermelon, pumpkin *(Cucurbita spp.)*, corn, bean *(Castanospermum austral)*, melon and sesame. The high numbers of uncultivated gardens (105 out of 185) correspond to people waiting for a tractor offered by the Mennonite organization ASCIM to plough their garden.

Figure 13. Crops cultivated in 80 home gardens

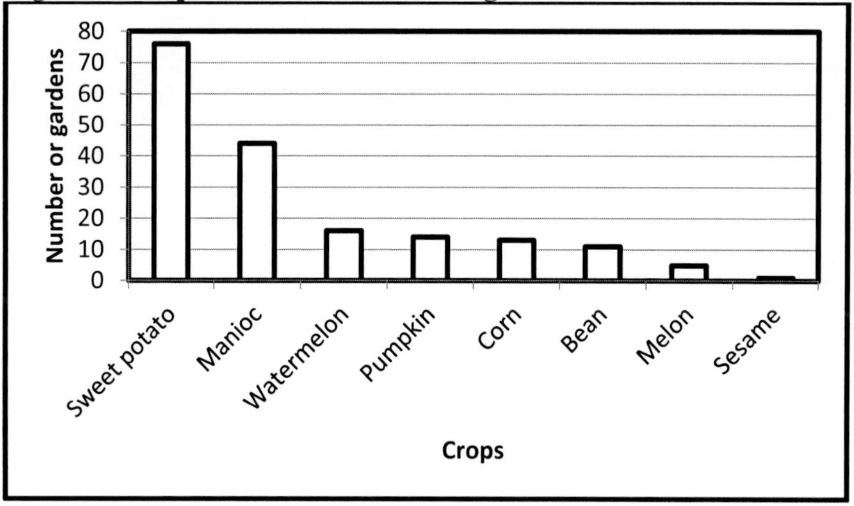

Mixed cropping (cultivation of two or more crops on the same plot), observed by Leake in 1995, is still practiced. This technique helps control pests and cope with rainfall variability. Of the total 185 gardens, 4,3% were cultivated with more than 3 crops, 26,4% were cultivated with two types of crop, and 28,6% with only one type.

In some villages, like *San Fernández*, *Karova Guasú*, and parts of *Laguna Hũ*, big and very productive orchards (mainly grapefruit) can be found. Most households in other villages constantly defend their garden and plants from the cattle, donkeys and horses that freely move within the colony.

In 2015, households seemed to depend from November to March on the crops they had produced. As crops are normally not sold and people do not practice any kind of storage, crops were harvested according to the daily needs. This makes it difficult to measure the overall yields. As it was observed, a sweet potato plot of 15 x 15 meters can be consumed by a household consisting of the parents and 6 children in a period of 4 months. In some cases, a garden produces much more than what is needed or can be consumed. In these cases, the yield is offered to other people. It is common to see women carrying bags full of watermelon, sweet potatoes or other crops from one village to another. People sometimes offer money or try to exchange for other things, like clothes. One man in *Karoa'í* village that had produced too much watermelon and was asked to sell some, commented *"I answered that I didn't want to do them any harm because we are all indigenous and offered them to take what they wanted."*[xix] I observed that not just "any" people asks for access to the yield and also not "only" relatives. A complex and constantly recreated network of relations and exchanges connects and disconnects the people of La Patria.

Some villages have experimented with bean and sesame cultivation at larger scales (1 ha) with the support of development projects or Paraguayans from *Colonia Ceibo*. The experience with beans was successful, allowing people to keep and share among members of their village and relatives in other villages. Yet, a second attempt failed because some villages received goats and sheep from a development project which they could not keep from destroying the plants. Some part of the harvest was also sold to or exchanged with the nearby *estancias* where some Angaité work. The opportunity to exchange products with Paraguayans seems to make people particularly happy. The sesame plantation generated some local work opportunities because the Paraguayan who was considered to "own" the plantation hired some men, paying per day. It is worth noting that despite the fact that the Angaité know that they are the owners of the land, the Paraguayan is considered to be the owner of the plantation because he asked the leader for permission and brought the seeds.

Development projects, like the ploughing service, that provide seeds and wire (to protect the gardens) from the ASCIM are appreciated by most of the villages. According to them, the only condition imposed by the ASCIM is

[xix] *"como ellos son todos compañeros indígenas, yo les dije que no quería perjudicarles y les dije que lleven nomas." Mr. Teodoro, June 2015, my emphasis*

that they have to sow and weed their gardens. If they fail to do so, they are excluded from the beneficiaries list the following year. On one hand, most households awaited the arrival of the tractor anxiously. Thus, they sometimes missed the opportunity to sow in a good moment because the tractor came too late. On the other, some people did not allow the tractor to plough their gardens anymore. They had observed that the clay soil became too hard after the second or third time it was ploughed. In most villages, bad experiences with communal plots caused the rejection of communal gardens. Despite the clear appreciation of this service provided by ASCIM, the trials I witnessed were not successful. In some villages like *Karoa'í, Tatarẽ, La Paciencia* and *Paraíso* absolutely nothing sowed after the ploughing germinated. Some people attributed this to the unusual heat and the subsequent heavy rain that affected the colony just after they sowed, at the end of October. Others said that the ploughing was done too late or with the wrong equipment. People in these villages did not have another provision of seeds to try a second attempt and just abandoned the idea of producing a harvest.

Due to some methodological differences, no direct comparison can be made to Leake's quantitative data set on home gardens. Nonetheless some changes can be mentioned. Firstly, the number of gardens that currently exist in the colony increased. This is probably related to the population growth experienced in the past 20 years and to the floods that resulted in a shortage of horticultural land. Secondly, the population now depends on external inputs, like seeds and tractors, which exposes them to uncontrollable risks. Finally, the climate variability that characterized the Chaco region in the last decade disrupted the "traditional" horticultural calendar. People implemented short cycle crops (2 to 4 months) as an adaptation strategy.

4.3.2.2 Vegetable gardens

A total of 37 vegetable gardens *(huertas)* were also recorded during the participatory mapping. Most of them were distributed among the four villages of *Paraíso, Monte Kué, Puente Kaigue* and *Urunde'y*. Vegetable gardens have been promoted by many development projects in La Patria. According to the respondents, the most important and effective training was conducted by the Agronomic Engineer Marlice Leguizamón within a project executed by the national a NGO. Leguizamón remembered that when she started the training in 1999 people did not show much interest. Her training methodology included long stays in the community, eating and sleeping most of the time there. This has become something very unusual and is highly appreciated by the residents. When she returned in 2015, as part of a Food Security Project executed by Grupo Sunu, she found that many families continued with the vegetable gardens. Some of them even saved their own seeds. However, most of the vegetable gardens identified in the participatory mapping were in the villages where Grupo Sunu was working in 2015. They supplied the households with seeds, a complete kit of tools and petrol or oil for chainsaws used to build fences. Female participants that took part in the training had asked for seeds of particular vegetable species that they didn't know how to eat. In many cases these were indeed not eaten. However, many women expressed how happy they felt when these unknown vegetables grew healthy and strong (*salieron bien).*

4.3.3 Livestock

In 1995, *San Fernández* was the only village with some cattle (about 50 heads). Most villages received animals from *La Herencia* project, run by the Anglicans as part of the re-settlement process, only a few years prior. According to Leake, most villages quickly sold or ate their cows. In addition, a rabies epidemic decimated the herd in 1993/4. He also pointed out that *"a delegated village foreman managed both private and communally owned herds (but) it was not possible to determine what recompense, if any, he obtained from this work."*[21]

During the participatory mapping exercise, I documented that six villages had cattle: *San Fernández, La Leona, Puente Kaigue, Colonia 24, Karoa'í* and *Karova Guasú.* The first five of this list got the cattle from a project of the INDI[xx] around 2009. *San Fernández* village received 9 cows and 1 bull. The herd grew to 28 animals in 5 years, but 4 were rustled. Absolutely no cattle were slaughtered by the villagers since they got them. This was mainly achieved due to the good relations that the leader maintains with the owners of neighbouring cattle ranches. Every now and then the ranches "donated" a cow to the village or La Patria colony when they had a special celebration. Only *San Fernández* village has a proper paddock, about 2000 hectares, built by the current leader's brother with the support of the Anglicans around 1987. Since then, some fences have been renewed. As the rest of the villages lack paddocks, the cattle move freely across La Patria, sometimes getting lost for a couple of days. There is constant tension with villages that lack animals and have their horticultural gardens damaged.

[xx] *INDI is the Instituto Paraguayo del Indígena. The name of the project was: "Improving the life conditions of indigenous people in their habitat".*

In each village cattle management is done in a slightly different way. In *San Fernández* the leader mixes the communal cattle with the private cattle owned by a Paraguayan that leases the village paddock. As the leader gets paid for taking care of the cattle that belongs to the Paraguayan, he is not getting any other recompense from the village. Each village has their own cattle brand, which is registered and recognized by the Paraguayan government. However, this brand is granted to individuals, most of the time they are village leaders or a relative. In most of the villages the leader's family or someone else designated by the leader is in charge of managing the cattle permanently. The leaders "own" the brand because they "got" the project and invest a considerable amount of time managing the cattle - naturally referred to as "my cattle". They are the ones who can at the last stage decide when to slaughter or sell an animal. Every time the cattle needs to be vaccinated, mainly to control the *aftosa* (foot and mouth disease), the leaders pay the cost. If leaders are not in a village when a decision has to be taken, the village assembly can make this decision instead. Cows are not milked in all villages. In *Puente Kaigue* the cattle basically live "with" the leader's family. Every afternoon, the cattle is locked next to the house. The docile cattle, in this case Hereford race, are kept watch over all night. In the other villages some cattle were eaten during extreme food shortage. No cattle are left in the *Seis de Marzo* village, mainly for the same reason.

A slightly different experience has taken place in *Karova Guasú,* conducted by a Mennonite organization. As it was explained by one of the leaders, three youngsters from this village ended their "biblical studies" with the Mennonites and as such they were entitled to get 8 cows, 1 bull and wire. The only condition was that they have to give back 8 animals in 5 years. In three years, the herd consisted of 7 cows, 1 bull and 6 calves.

During my fieldwork, cattle raising was again promoted by a project conducted jointly by the Paraguayan Government and UNDP. Most of the villages, again, showed to be interested in having cattle. As a starting measure, villages were offered wire to build their own paddock fences and food to work. As fences were historically the way private property was reinforced in the Chaco and as the limits among the different villages was not as clear as expected, this generated a series of tensions considered in Chapter 6.

As aforementioned, there was a failed cattle experience as part of *La Herencia* project in the late 80s. According to an Anglican technician that I interviewed the main problem was the way the cattle was handed to the Angaité people. First, they only worked with people from two villages who showed to have real interest and where two paddocks were built with some supports from the Anglicans. In 1987, the project manager decided to hand over all the cattle to the indigenous people. That year, 700 heads of cattle were distributed among the villages of La Patria! He mentioned two main reasons why that caused the failure. The first one is that *"the Angaité were not ready yet"*[xxi] and the second is that it generated envy among those people who got less cattle or none.

Many times, I asked the Angaité people and myself: Why is there so little cattle in La Patria? This is perplexing because they have around one century of working experience in cattle ranches. Mr. Santiago, leader of *Karoa'í* village, referred to the failed experience with the Anglicans in his answer. He explained that a strong frost ruined all the gardens and the only available resource then were the newly received cattle. He also mentioned that *"it is only now that there is a cattle (development) project again"*. It is worth noting that the interviewee attributes a passive position to himself in reference to external projects. He is the "receiver" or "beneficiary" and perceives himself as unable to start or work towards their own ideas or plans. Young leaders also refer to this failed project and accuse the leaders of those years of being incompetent or "asleep".

Villagra analyses an Angaité myth about the original loss of the cattle. He states that the cultural importance of cattle *"derives from their high potential as a means of maximizing commensality and, therefore, of making possible the desire state of contentment and tranquillity."*[22] However, he also suggests that this potential is always ambivalent because *"it can equally provoke the negative antithesis: namely, internal dispute within villages and interethnic conflict between indigenous people and Paraguayans."*[23]

Besides the cattle, some villages also have sheep, goats, donkeys and chickens. In *Tataré, Laguna Teja, Carpincho* and *Monte Kué* villages, sheep and goat are a very important resource during long flood season. In most cases livestock is also managed by the leader, using the ruins of failed communal lavatories projects as barns. Donkeys are still the cheapest and most commonly owned form of transport despite the increased number of motorcycles (about 2 to 4 per village). As Leake observed 20 years ago, donkeys roam freely within the colony. Donkeys are owned by individuals and conflicts regarding ownership are not frequent despite the fact that donkeys from different villages mix up regularly. The donkey population

[xxi] *Mr. Vicente, personal communication.*

decimated around year 2010, when ranch owners paid a relatively good amount of money for them in order to use donkeys as lure to avoid jaguars from killing cattle on their ranches.

4.3.4 Land leases for cattle

The land leases to Paraguayans from the nearby town of *Colonia Ceibo* or to some neighbouring ranchers started a few years ago in La Patria. This activity is explicitly prohibited by the national law. However, neither the police nor any other institution enforces the law to stop it. This situation is also common in other indigenous communities in Paraguay. After the conflicts generated by the last land lease attempt of the former national senator Quiñonez, who owns *estancia Mirian Victoria*, located to the West border of La Patria, there are only two current land leases. The first and most important one takes place in *San Fernández* village, and it is related to the fact that this village is the only one that has a proper paddock. According to the information I got about this sensitive issue, *San Fernández* leader receives 1,500,000 G. (244,18 €)[xxii]every month to allow a Paraguayan from *Colonia Ceibo* to keep around 360 head of cattle in the village paddock. He manages the animals with the support of other men from the village who get paid by the Paraguayan. However, over the past years people from *San Fernández* and other villages have increasingly resisted this agreement. This resistance is expressed in different ways, for example by politically excluding the main leader or even "hunting" some cattle (more about this conflict, see Chapter 6). Besides this permanent agreement, there is a more occasional type of land lease going on in *Colonia 24, Tataré* and *Laguna Teja* villages. The neighbouring ranch, called *estancia Buena Vista*, pays to bring their cattle to La Patria colony when they run out of pastures. Leaders from this villages charge the amount of 100,000 G. (16,27 €) per day. The last time it happened, the cattle stayed 15 days in the areas that correspond to each village.

4.3.5 Timber extraction within La Patria colony

Cutting and selling *quebracho* post within the colony is another way of making a living. Paraguayan men involved in this activity are locally known as *posteros.* This activity is not new, Leake observed that the Angaité occasionally cut and

[xxii] *Exchange rate of 01.09.2016 (1Eu = 6,143 G)*

sold *quebracho* and palm posts due to the increasing local demand of the surrounding cattle ranches. He also mentioned that *posteros* were arranging contracts with villages and that *"villages could only outpost within their allotted area of the colony."*[24] During his fieldwork periods from 2004 to 2006, Rodrigo Villagra also observed in detail what kind of relations and transactions generated this activity in the colony between the Angaité and the Paraguayans.

"Ñamba'apo comunidad ryepype" ("we work inside the colony") is how the Angaité affirm that someone or a village is involved in timber extraction. "Work" does not normally refer to agriculture, hunting, gathering or fishing. "Work" usually refers to activities that involve external *patrones.* According to local leaders, there was a considerable increase of timber extraction since the year 2000, and also more recently, considering that more people own chainsaws. According to my survey in *Karoa'í* village, 5 (18,5%) men out of 27, were involved in timber extraction during 2015. Dry weather conditions are favourable, as they are for many other subsistence strategies, because they facilitate transport on the colony's complicated internal tracks.

The *posteros* or loggers are all Paraguayans, some of them are fully dedicated to this activity, and others combine it with other work. Some loggers are also *macateros.* Others have a formal job at the nearby *estancias* and get involved in this activity to generate an extra income. There are two distinct ways these loggers operate. Most of the time loggers make a deal with the village leaders. In 2015, a Paraguayan from *Colonia Ceibo* known as *Gordo,* had this kind of arrangements with leaders from at least 8 villages: *Laguna Teja, Laguna Hũ, Monte Kué, Carpincho, Puente Kaigue, Paraíso, Karoa'í, Las Flores* and *Tatarẽ. Laguna Teja* was his main source of *quebracho* the previous years but the heavy rainfalls of 2014 and 2015 forced him to establish new deals with other villages. Depending on the sizes, one *quebracho* tree can turn into a smooth-post (*poste liso)* or into 4 to 8 cracked-posts (*poste rajado).* This *postero* and Mr. Diego, leader of *Tatarẽ* village, openly shared the terms of their agreement. To give his permission to cut off *quebracho* in *Tatarẽ* village area, Mr. Diego gets 4,000,000 G. (651.14 €[xxiii]) for each semi-trailer truck (about 700 *quebracho liso*-post). The *postero* organizes the whole extraction process in addition to providing chainsaws, oil, petrol, and food for those involved. Men from different villages that work for the *postero* disregard the village-area where the *quebracho* is cut off. To the men using chainsaws, he pays 60,000

[xxiii] *Exchange rate of 01.09.2016 (1Eu = 6,143 G)*

G. (9.76 €) per day and provides food (*libre*). The others, who work with axes and machetes, in charge of pealing the quebracho bark, get paid 11,000 G. (1.79 €) per post. Other loggers had similar arrangements with other village leaders, sometimes paying them in cash or in kind. As part of their deal *Karoa'í* leader received 30 kilos of meat from a logger who also works as a mechanic in a nearby *estancia*. The first *postero* mentioned estimated that he worked with 70 households in 2015. He assured me that he is "*helping the indigenous people a lot, giving them a job to satisfy their needs.*"[xxiv] He also stated, *"I help them a lot, and I pay. The other (loggers) are fooling them because they don't read."*[xxv]

In the last years, however, some people eluded the leader's intermediation and started to sell post directly to the loggers, with prices of approximately 40,000 G. (6.51 €) for smooth posts and 12,500 G. (2.03 €) for cracked-post. A man from *Karoa'í* explained why he sold post to a specific logger, not in terms of a better deal, but just "*because he is always nice to us.*"[xxvi]

As said, this activity is conducted illegally. As loggers don't have the official permits issued by the *Instituto Forestal Nacional,* there is no monitoring or record of the extraction rate in the past years. One logger estimated that around 7,000 smooth-posts have been extracted in the last 2 years just by himself. According to six other respondents, there seems to be very little *quebracho* left in the easily accessible forest of the colony. Leaders and other people know that it is forbidden by law to use the forest resources this way. They normally justify this activity by saying that they do it *"because they have needs"*, because *"there are no alternatives"* or *"because the emergency aid is not coming so often."*[xxvii] Some leaders expressed that they would like to get the official concession from the government, *"but the government says it is illegal for indigenous communities, and it seems that the government, as anthropologist, want to keep us living as we live."*[xxviii]

[xxiv] *"Mucho estoy ayudando a los indígenas, porque les doy trabajo para salvar sus necesidades".*

[xxv] *"Yo les ayudo mucho y le pago, otros les joden, como no saben leer."* Mr. Gordo, September 2015

[xxvi] *"Ore rome'ẽ Adriánope, porque ha'e ikalida iterei ore ndive."* Mr. R. Riquelme, July 2015

[xxvii] Mr. Pablo, September 2015

[xxviii] *"... el gobierno no permite que los indigenas vendan su madera. Tampoco les permiten que alquilen, parece que el gobierno, como los antropólogos, quiere que sigamos viviendo así mismo como vivimos."* Mr. Pablo, September 2015

4.3.6 Wage labour

In 1995, half of the households of *San Fernández* village had members who periodically sought wage labour on ranches.[25] According to the duration of their work, they are locally classified as *mensualeros* and *pordiaseros*. The first refers to permanent employment and the latter to occasional employment. My survey in the same village found an apparent decrease. Out of a sample of 48 men, 18 (37.5%) worked on cattle ranches. Out of these, 13 (27%) did so as permanent workers and 5 (10.4%) as occasional workers.[xxix] However, it has to be said that during 2015 a considerable number of men decided instead to work for the loggers inside the colony. The same survey was applied in *Karoa'í* village to a sample of 27 men, where 19 (70%) mentioned that they worked on ranches, 13 (48%) as permanent workers and 6 (22%) as occasional workers. Besides these, other 5 men (18.5%) mentioned that they were working for the loggers, occasionally combined with wage labour on ranches. So, in *Karoa'í,* the percentage of men involved in wage labour is as high as 78.5%!

According to the 2012 Paraguayan Indigenous Population Census, 48.3% of La Patria population is economically active, most of them in the "agriculture and livestock" sector.[26] Unfortunately, this Census does not differentiate if the activity refers to agriculture and livestock inside the community or wage labour on cattle ranches. Some leaders commented that people were working less in the *estancias* in recent years, as a result of the increased development projects and the possibility of "working in the community" selling *quebracho* posts to loggers.

On the *estancias,* indigenous workers occupy the following lower ranking positions: *retirero,* the person in charge of an outpost attached to a particular sector of the *estancia; estanciero* or "cowboy"; *playero,* a person who performs tasks on foot, such as clearing the grazing fields of trunks and mending fences.[27] As in 1995, the type of work undertaken on ranches depended on particular skills. Most people obtained manual tasks as *playeros* related to the maintenance of ranch infrastructure, clearing of thorn bush from pastures, planting grass and repairing or building fences. These activities depend on the weather. In 2015, people who had these kinds of temporary jobs worked from May to October. Women are less likely to be hired to work on a ranch. The ones in charge of cooking for the personnel are called *machu.* In most of

[xxix] *I excluded women because only a very reduced number work in ranches.*

the cases, however, female work is included in the husband's wage. In the Mennonites area, indigenous women or their children occasionally work as house cleaners. Apparently, young couples tend to work far away for a couple of years or until their children turn five and then move back to the colony. The following higher ranking positions are mainly occupied by Paraguayans: *patron,* is the owner of the *estancia*; *mayordomo*, the foreman; and *contratista*, a person hired to build fences, corrals and houses.[28] In La Patria, "Powerful outsiders" - such as the *patrones* (as well as missionaries and anthropologists) - are referred to as *veske*[xxx]. The latter is the term the Angaité used to refer to their own leaders in the past.[29]

Rates of pay vary depending on the provision or lack of food while working at a ranch. If some provisions are included, the deal is referred to as *libre* (free). If no provisions are included it is referred to as *seco* (dry). In 2015, the rates of pay at nearby ranches were on average of 37,000 G. (6.02 €[xxxi]) per day and the kind of payment was *libre,* it included provisions. The rates were relatively higher in the Mennonites colonies' area, ranch workers earned 55,000 G. (8.95 €) per day and also had a *libre* deal. Some men permanently working in ranches owned by Mennonites also have AMH, a mixed private-public health insurance system.[xxxii] Food provided by the *patrones,* generally speaking, barely covered the quantity and quality minimum the workers require. Additional provisions are sold in the ranches at high prices to be discounted from future payments. The better rate paid in the Mennonites colonies area is countered by the higher cost of extra provisions and higher cost of living and transport in the area. Another type of arrangement, locally known as *por tanto,* refers to a deal made with a single person, called *cabezante*, hired to fulfil a specific task for a fixed price negotiated beforehand. Only the *cabezante* is responsible to the *patron* and has complete freedom to arrange payment conditions with the people he hires. The latter are usually, but not only, his relatives.

Working conditions and relations are expressed in affective terms. *"We don't know if our patron is going to love us forever,"*[xxxiii] expressed a local leader when describing such relations. Others said that their *patrones* were *"good people"* or

[xxx] *In Guaraní, ranchowners and loggers are referred to as che patron, che jefe or che ruvicha, "my patron", "my boss" or "my superior". Villagra, 2010*

[xxxi] *Exchange rate of 01.09.2016 (1Eu = 6,143 G)*

[xxxii] *In 2012, 9,90% of the total population of La Patria had any kind of health insurance. DGEEC, 2012*

[xxxiii] *"...ndajaikuaái la ñande patrónpa ñande rayhúta todo el tiempo."*

"nice people."[xxxiv] Villagra states that, as the patrones are referred as *veske* (term formerly used to refer to Angaité leaders), they are *"expected as such to be as generous as the former"* and that what lies beneath labour relationship are not merely formal obligations *"but the good faith and care shown by a mutual desire to help each other."*[30]

Paid work is conducted in at least 19 different cattle ranches. Some of them located near the colony and others considerably far, in the Mennonites colonies of Loma Plata, Filadelfia and Neuland. Respondents mentioned that they work in *estancias* or areas that they consider as their place of origin. Mr. Santiago stated: *"Right now there are young people from Karoa'í working in Estancia Carmelitas. The people do not forget their community, because their grandfather grew up and lived there, and their parents. And they know it. And they go there now. They go to work and to hunt, gather and fish. People from Karoa'í, Paraíso, and Seis de Marzo. Because they are all descendants from Karova Guasú village."* [xxxv]

4.3.7 State and NGOs related income

In 2015, 3 main governmental programs were implemented in La Patria colony and translated into direct monetary benefits or food provisions. The first one is *Tekoporã,* this program depends on the *Secretaría de Acción Social. Tekoporã* targets households that live in extreme poverty with at least one child from 0 to 18 years old, a handicapped member or a pregnant woman. It is classified as a Conditional Cash Transfer program which provides every family with 450,000 G. (73.25 €[xxxvi]) every second month and an annual total benefit of 2,700,000 G. (439 €). Some families from La Patria were included as beneficiaries of this program around the year 2009. In 2012, 39 households (10%) of La Patria were covered.[31] According to my observations and the leader's estimations 660 "families" participated in this program, in 2015. This includes almost the whole population of La Patria.[xxxvii] The money is paid in cash to women in a spectacular event that gathers most of the population of

[xxxiv] *"buena gente son, calidad son"*

[xxxv] *"Ko'a pegua, ko'ãgaite oiméne oho mitãrusu Carmelitaspe (…) la hekoha ndahesarái porque upépe voi oiko, okakuaa va'ekue la É, itúva kuéra, pero la familiare (descendientes) oikuaa oiko va'ekue upépe, michi rire. Ha upépente oho ko'ãga. Oho jey ochanga, omariska. Karoaigua, entre Paraisogua. Porque péva KG familiare. Oĩva Paraiso ha 6M"*

[xxxvi] *Exchange rate of 01.09.2016 (1€ = 6,143 G)*

[xxxvii] *According to official sources, at a national level, 66% of the indigenous communities are beneficiaries of the Tekoporä program. This represents 16.500 families. (http://www.sas.gov.py/noticia/1878-cobertura-de-tekopor-llega-al-66-de-los-indigenas.html#.V9pMlK355-8)*

La Patria in 2 villages, *La Leona* and *Paraíso* (see picture 18). Due to access conditions, payment sometimes gets delayed a few weeks or months. The announcement was made via the *Pa'i Puku* radio, the most important of the region, because there was no telephone communication in the area. Sometimes payment was distributed earlier because expected people from all villages of La Patria gathered at least one day before and waited a couple of days until the military protected convoy arrived into the colony. A convoy integrated by *macateros* arrives either before or after the military one.

The second program is a maintenance pension for elders (*Pensión alimentaria para adultos mayores)*. It targets people living in poverty conditions that are 65 or older and depends on the Ministry of Economic Affairs (*Ministerio de Hacienda)*. Beneficiaries receive 456,000 G. (74.23 Eu) monthly, 25% of the national minimum wage.[32] In 2012, only 19 elders (4.9%) of La Patria were covered. However, this number increased at least by three in the past years. Payment is done via a debit card, which constitutes a real problem for the elders who normally don't travel out of the colony. Most of the beneficiaries make an agreement with the *macateros*, who take their debit cards to the city and then provide them with supplies. Informally, some *macateros* manage the elder's money and only sometimes giving them a portion of the cash. These elders are locally called *tarjeteros* (those who have a card). In scarcity conditions or when the *macateros* cannot reach the colony easily, these elders are the first to get supplies.

The third program is an emergency aid plan conducted by the National Secretary for Emergencies (*Secretaría de Emergencia Nacional)*. Since the floods in 2014, when La Patria colony received emergency kits on 3 occasions, households got used to receiving aid and are somehow constantly expecting "the emergency" to occur. The "family census," used to register the beneficiaries, reorganises the members of each village into "families". In some villages the relatives of the leaders were registered as many "families" as a strategy to get more food kits.[xxxviii] During my fieldwork the colony received these kits on 3 occasions (see picture 17). The role of the village leaders was to register families and elders in the governmental programs and

[xxxviii] *I witnessed one Emergency kit distribution. Every kit consisted in: rice (5 Kg), flour (5 Kg), sugar (5 Kg), yerba mate (5 Kg), pasta (3 Kg), oil (5 Kg), salt (1 Kg), peanut (5 Kg), beans (5 Kg), biscuits (5 Kg), canned beef (2 units), soap (2 units). After a few days people threw away the yerba mate because they said it was causing them stomach-ache. Every kit has an estimate caloric value of 161,147 Kcal, and a monetary value of 272,250 G.*

to request emergency aid in cases of extreme weather conditions, both duties are vital to get the benefit of these programs.

Besides these programs, there was a temporal and thematic superposition of development projects conducted by governmental institutions or NGOs in the colony. On certain occasions, different institutions organized communal meetings at the same time and leaders had to act strategically to try to "get something" from the external visitors. Leaders expressed their interests in almost anything proposed by the visitors. On my last visit to *Puente Kaigue* village, the leader commented that they were building or planning: fences for a paddock, a beehive production project, a new water reservoir for fish farming, a communal bakery for the women, a goat breading initiative and vegetable gardens. All these projects were promoted by different NGOs more or less at the same time. The expression used by the leader to describe their success in getting all these projects was *roagarrapá,* it literarily means "we got/catch it all". External benefits are generically called *ayuda*, which can be translated as "help". Nevertheless, there could be a real need or interest in experimenting and getting to know/learn from all these projects. This complete openness and simultaneous interest suggests a concrete concern for the direct and immediate benefits that these projects can provide. Another real concern is to establish social relations with external people and institutions to which they could appeal if they need some help.

4.4 Conclusion

The livelihood strategies of the people of La Patria colony consist of a broad portfolio of activities: paid labour, livestock raising, beekeeping, horticulture, hunting, gathering and fishing, timber extraction and land leases. The specific strategy and portfolio of each household depends on the development cycle of the domestic group, the year season, the weather conditions and their access to equipment as well as resources needed for each activity. I found that the majority of the households were equipped to engage in most of the activities, excluding beekeeping and land leasing. Most of the livelihood activities involve people from more than one household. Therefore, households from each village are interrelated and interdependent through a series of relations and transactions that are not restricted to kin or relatives. Besides the fact that some households are less interdependent than others, the village level approach is the most suitable for understanding the dynamics related to livelihood strategies in their full complexity.

Due to the decreased number of wild animals inside the colony, hunting currently requires more effort than in 1995. On one hand, this decrease is caused by the augmenting pressure of a growing population. On the other, it is caused by the deforestation process occurring in the region in the last years. Hunting outside the colony limits has also decreased comparing to 1995. It is due to the deforestation but also to the increasing and illegal restrictions on private properties access by the ranch owners. Even hunting in their spare time while working or visiting relatives in cattle ranches has been denied in the last years.

The access loss to private properties also affected fishing and gathering as subsistence activities. Despite the fact that households are depending less on fish than in 1995, it is still a very important source of food. Wild honey is not only an important source of calories but also an important source of money in the summer, when wage labour in cattle ranches decreases due to the summer rains and the extreme hot temperatures. Despite the apparent decreased importance of wild fruits and plants in the current diet, these are critical for survival during floods or other kinds of food scarcity situations.

The higher number of gardens that currently exist in the colony, probably related to the population growth experienced over the past 20 years, results in lesser availability of horticultural land. The colony population is now depending on external inputs, like seeds and tractors, which expose them to uncontrollable risks. The weather variability that characterized the Chaco region in the past decade disrupted the traditional horticultural cycle and calendar. People adapted to this variability by tending to prefer short cycle crops (2 to 4 months).

Wage work at the cattle ranches is still an important activity for the majority of the households and villages. In some villages, the percentage of men involved in wage labour is as high as 78.5%. On the *estancias* indigenous workers are still occupying the lower ranks. Rates of pay vary according to the location of the ranches and the specific agreement with the *patrones.* There is an increased number of people working in the Mennonite area, in slightly better conditions. The Angaité describe working conditions in affective terms rather than as a formal relationship. Hunting while working on cattle ranches has also decreased in the last years.

The Angaité livelihood activities, as suggested by Ingold for other hunters and gatherers, involve management as well as care and knowledge. In fact, care and help seem to be central to how the Angaité conceive the social relations in which they get involved through their livelihood activities.

The notion of procurement, suggested by Ingold, solves the tension between the notion of "foraging" as used by Leake and "production" as used by Gordillo. I mentioned in the previous chapter that the Angaité conceive "La Patria territory" as a place not only inhabited by themselves, the animals, plants and other non-humans but also by the ranchers as "not so human" others. Despite the changes described in this chapter, there are signs that this particular understanding of the "social collective" that inhabits "La Patria territory" persists. The fact that the Angaité mentioned, during the participatory mapping, the *Jary Ruvicha*, a "being" that protects animals; the shaman that talks to fishes; and the *algarrobo* season that never reached again are direct signs that this understanding persists. The manner in which the Angaité describe their relations with the ranchers or *patrones* in terms of "love", "care", "being nice" or "help", is also another indication. A less evident but important sign is the interest they show in learning about things they do not master yet, including but not restricted to their interest in growing plants they don't eat and their involvement in any kind of external projects. Paraphrasing Blaser, the indigenous people learned not to mention those aspects of their ontologies that could be considered, from a modern point of view, as "not serious" (or even ridiculous).[33] Thus, these aforementioned signs should be considered the top of the iceberg, the openly showed and verbalized aspect of the persistent Angaité cosmography. This conscious omission of the role played by non-human beings and shamans was also mentioned by Bonifacio and Villagra.[34]

In short, the inhabitants of La Patria colony are depending less on hunting, fishing, gathering and agriculture than 20 years ago. They are, however, more dependent on wage labour from cattle ranches, development projects and state programs like *Tekoporã,* the emergency aid and the pensions for elders. They are also increasingly involved in cattle raising and timber extraction. Some villages are likewise generating some economic resources through land leases for cattle. Cattle raising, timber extraction and land leases involve the use of communal resources inside the colony. As there are not clear village limits or any official subdivision of the colony, there are an increased number of conflicts, tensions and challenges related to these activities.

Leake stated that *"the evidence points to the fact that where given the choice, [the Angaité] will continue to rely on foraging as an important, if not total, basis for subsistence."*[35] Therefore, it can be claimed that the Angaité were forced to modify their livelihood strategies by the destruction of and access restriction to resources on which they depended. The state had an active role in this process. The state forced the Angaité to modify their livelihood strategies through active interventions and development projects of the state and NGOs as well as the *de facto* not recognition of the Angaité territorial rights when approving the RIMAs.

Despite all the environmental and sociocultural changes in the region over the past 20 years, the inhabitants of La Patria modified their livelihood strategies as little as necessary in order to maintain their own particular form of social organization. When Kidd analysed how the increasing dependence on wage labour impacted the Enxet society, he concluded that the *"tremendous transformation in their economy… is, in one fundamental aspect, more apparent than real."*[36] He highlighted the fact that as in pre-colonial times, the Enxet continued to practice a diversified economy, which is perfectly compatible with their moral values involved in the practice of immediate consumption. In the case of the Angaité, this makes sense considering the relation between the autonomy and freedom of these individuals and the kind of activities they get involved in throughout the year or their lives. However, as I show in the next chapters, the abandonment of some livelihood strategies implies the cease of some processes of *enskilment* or "training of attention" through which values and cultural knowledge are transmitted from one generation to the other. In addition, the increasing circulation of money, obtained through wage labour, timber extraction, land leases or state assistence programs, seems to have an impact on the Angaité's forms of sharing. The changes experienced by the Angaité are not only "economic." The increased relation with and dependence on the state also implied important transformations of what is usually considered as the "political" domain. This is why, in the next chapter, I focus on the relations with the state and other patronage networks.

---

[1] Tim Ingold, 2000: 58
[2] Ibid., 133
[3] Marschke, M. J., and F. Berkes, 2006
[4] Tim Ingold, 2000
[5] Ibid., 59
[6] Volker Von Bremen, 1987: 12
[7] Nurit Bird-David, 1990
[8] Volker Von Bremen, 1987
[9] Gastón Gordillo, 1993: 91
[10] Andrew Leake, 1998
[11] DGEEC, 2014

[12] Ibid.
[13] Andrew Leake, 1998
[14] DGEEC, 2014
[15] Andrew Leake, 1998
[16] DGEEC, 2014
[17] Ibid.
[18] Ibid.
[19] Andrew Leake, 1998: 155
[20] Ibid.
[21] Ibid.
[22] Rodrigo Villagra, 2010
[23] Ibid.
[24] Andrew Leake, 1998
[25] Ibid.
[26] DGEEC, 2015
[27] Rodrigo Villagra, 2010
[28] Ibid.
[29] Ibid.
[30] Ibid.
[31] DGEEC, 2014
[32] http://www.hacienda.gov.py/web-sseaf/index.php?c=181
[33] Mario Blaser, 2010
[34] Valentina Bonifacio and Rodrigo Villagra, 2016
[35] Andrew Leake, 1998: 257
[36] Stephen W. Kidd, 2000b: 141

5 Development, patronage networks and the "relational perspective"

5.1 Introduction

In this chapter, I describe and analyse some specific relations of the Angaité of La Patria with external actors (missionaries, state and NGOs). This issue has been researched by several anthropologists specialised in the Chaco region and is central to ongoing discussions about indigenous territorial management, territorial rights and political protagonism. Land and territory, in this context, is not only considered an essential basis for economic development but, in Leake's words, "*as a means of enabling [the Angaité] to recuperate lost psychological space to develop their own society in their own criteria.*"[1] The changes in access to and the use and management of natural resources took place at the same time than the transition from the Anglican patronage to the state/NGO patronage. The responses of the Angaité to each of these phenomena cannot be easily separated. In this chapter, I try to show the impact of two parallel processes: the parallel reduction and expansion of the limits of the social collective. To do so, I will explore how these two processes reflect on the changes suffered by the Angaité leadership practices and forms of sharing.

The relations of the Angaité with external *patrones* are part of 'development', understood as a governmental practice and discourse. For decades now, some scholars have been approaching development from a critical and postcolonial perspective. Thus, revealing development's aim "*of (re)creating modernity and its institutions elsewhere, under the guise of civilizing mission, colonial rule or development proper.*"[2] In this context, central concepts of the development discourse – including: "help", "science", "development", "market", "needs", "poverty", "participation", "production" and "resources" – have been deconstructed to reveal their role to consolidate a Western vision of reality.[3]

Gustavo Esteva's work contributes to this critique with a series of elements to rethink "development". He argues that the "development era" started with Harry S. Truman presidential speech in 1949. In that opportunity, Truman stated: "*We must embark on a bold new program for making the benefits of our scientific advances and industrial progress available for the improvement and growth of underdeveloped areas. The old imperialism -exploitation for foreign profit- has no place in our plans. What we envisage is a program of development based on the concepts of democratic fair dealing.*"[4] He called all the people and societies that did not base their life on the industrial mode of production "underdeveloped." Truman gave a new political meaning and use to the modern notion of "development". This modern notion refers to the "*the active principle according to which new and higher stages of human society might emerge out of old and more simple ones.*"[5]

The modern notion of "development" resulted from a process of at least two centuries of social construction and is closely related to the Western nature-society divide. In the 18th century, the relation of nature and society started to be framed by the Western Europe writing elite in a temporal matrix: nature was the starting point of humankind's voyage towards the future and progress was conceived as the progressive domination of humankind over nature.[6] Studying concrete cases of development practice and discourse, scholars like Ferguson argue, that development's instrumental effect is the extension of the state bureaucratic power.[7] Others counter-argue, scholars like Blaser, that in contexts like Paraguay, "*development extended a form of power which was expressed through an all-pervasive patronage system.*"[8] He claims that the extension of the patronage networks was not an ´anomaly´ in relation to a normal model of development, rather it was the concrete way in which "development" materialized as practice in this particular setting.[9]

I mentioned before that Von Bremen[10] contended that the indigenous people's conduct towards the development projects was determined by specific conceptions, which traditional hunter gatherer lifestyle were based on, and that these were still latent despite the long history of domination. How, then, were these conceptions and values determining indigenous people's responses when Von Bremen wrote his report in 1987? According to this argument, based on the same logic of the traditional relations with nature, "*indigenous people are trying to know more about the projects and to adapt to it characteristics…*"[11] The indigenous people try, in this context, to make the whites (missionaries, anthropologist, etc.) share their knowledge of new phenomena.[12] Following this argument, the indigenous leaders related to development projects as traditional leaders (mainly shamans) related to the spirits in order to appropriate the powers and resources needed to live well.

As mentioned before, Blaser situated Von Bremen (as well as other anthropologist like Renshaw and Kidd) within the group of "development experts" that shape the regime of truth of modernity. He argued that as such, Von Bremen was heavily imbued with the Cartesian moral logic and as a result tended to translate practices embodying relational moral logics into immutable traits.[13] As an anthropologist assessing development projects, Blaser says, Von Bremen was exposed to mutually contesting demands - on one side, the need for clear-cut categories, and on the other, the commitment to cultural diversity. One of Blaser's critiques to Von Bremen was his suggestion that development projects appear to the indigenous people as "*(modern) hunting and gathering grounds where resources abound.*"[14] According to Von Bremen, the validity of the hunter-gatherer mentality was evidenced by the shifting alliances of indigenous people with competing projects. Blaser counterargues that the shifting alliances have to do with an all-pervasive patronage system and not with any particular hunter-gatherer mentality. In order to refute Von Bremen's reasoning regarding the persistence of this mentality, Blaser states that "*by the 1980s most of these peoples were subsisting on the basis of a mixed economy where wage-labour and several forms of agriculture were very important.*"[15] With this quote, Blaser seems to argue that hunting and gathering was something of the past; therefore, not relevant enough to explain contemporary indigenous responses. In this sense, Blaser concludes that Von Bremen implicitly assumes that indigenous moralities remain the same throughout history. I conclude this chapter by arguing that my ethnographic data and that of other scholars suggest that Von Bremen's argument, regarding the persistence of a particular perspective and responses, might still be useful to understanding the Angaité responses. I also show that this persistence neither denies the influence of the patronage networks nor other historical transformations.

In the following section, I describe some specific patronage networks. In doing so, I show that a particular way to relate with their territory and external actors persists. This indicates that the current Angaité leaders keep relating to external actors the way shamans related to non-human beings. It is as though these relations are not ontologically divergent.

5.2 Patronage networks and the persistence of the "relational perspective"

The first time I visited La Patria, I was surprised by the visible remaining artefacts of development projects that were in disuse or used for a different purpose. Broken water tanks were used as shelters within which they made fires and cooked. Iron structures of the failed state housing project were used to create open kitchen's roofs. Irrigation systems were used to store fresh potable water. Unfinished houses were used as barns. Communal sinks and showers were used as residences. These are just some examples of what I mean by artefacts. However, and this must be clear, some projects had clearly been successful and vital to the Angaité's subsistence - such as the water tanks to store rain water, the communal carts to collect wood, the horticultural gardens' fences. I do not address the reasons for the development projects' failures here. Instead, I focus on how these development interventions and patronage networks impacted the Angaité cosmography.

Regarding the influence of development projects and patronage networks on indigenous forms of leadership in the Chaco, Kalish argued that a serious crisis was taking place.[16] According to him, the crisis was caused by the imposition of Western representative leadership mechanisms on the indigenous communities. These mechanisms, he argues, were not functioning inside the communities; within the community the leaders were just expected to be counsellors. Meanwhile, traditional leadership mechanisms were losing validity because they were relegated by the new leaders. If new leaders wanted to give some space to the traditional leaders/shamans, they would face the limitations of the imposed system.[17]

Villagra also explored the transformation of the *Enlhet-Enenlhet's* leadership.[18] He argues that despite its constant transformations some features of the past are still valid in the present. Villagra includes a series of personal abilities or qualities among these features, such as: non-coercive consensual authority; extensive kinship ties or relatedness represented by the number of followers; persuasive speaking (including knowledge of foreign languages); care, generosity and initiative; and courage in warfare and shamanic powers.[19] He argues that in the past the role of shamans and leaders were combined in the figure of the shaman-leader. In the present, as we will see, these two roles tend increasingly to diverge. If we consider that both shamans and leaders role's presently involve "*transformative skills in order to gain the generative power of 'others'*", the divergence of current shamans and leaders may not be that great.[20]

Villagra also suggests that the Anglican efforts to Christianize the Enxet and the Angaité were interpreted, by the latter, as proof of the Anglican's shamanic attributes. Anglican missionaries' behaviour and teachings fortified the conviction that they were shamans with their own techniques and auxiliary spirits.[21] Anglican distribution of wealth (as trade or gift), the simultaneous spreading of epidemics (sign of their healing or evil powers), the Anglican intermediary role with the other *patrones* (like the IPC company) and the shelter provided during the Chaco War were, according to Villagra, among the several circumstances that boosted this interpretation.[22] The Angaité realized that their *veske* (leaders), shamans, were no longer able to protect them and started to look for protectors among the non-indigenous *veske,* missionaries and ranchers.[23] Decades later, with the decreasing influence of the Anglican missionaries countered by the increasing presence of the Paraguayan state and its laws, anthropologists and lawyers were perceived to be as powerful *veske or* shaman-leader.

5.3 The new *patrones* and the redefinition of the "political"

The Paraguayan state relied on the missionaries and militaries to fulfil its presence in the Chaco since the beginning of its colonization. In the case of the Angaité of La Patria, the decreasing presence and influence of the Anglicans, since 1987, initiated a new era: the state driven development era. Since 1975, the Paraguayan law established procedures allowing indigenous people to make land claims. However, due to the influence of landowners and the longstanding military dictatorship, indigenous peoples only managed to secure small portions of land. As mentioned before, in those years, the Paraguayan government's declared aim was to 'integrate' the indigenous peoples into the national society.

According to the Angaité, the turning point for the state driven development era was in 1989 when their leaders, formerly organised in the (Anglican) Indian Directive Board, were officially recognised by the Paraguayan state. A second indicator of the increasing influence of state patronage was when the Angaité received La Patria colony land property title, in 1995. According to what they say, thanks to the title, many development projects started to "arrive". However, there was a transitional phase in the substitution of the Anglicans presence for that of the state. Another external actor emerged during this phase, the NGOs, appeared to intermediate between the state and the Angaité. As the missionaries were representation of the Anglican church, the lawyers and anthropologist working for the NGOs appeared as new powerful *patrones*. The Angaité of La Patria remember this transition from the missionaries to the state/NGO era. The first and most influential NGO in the area, TierraViva, supported (and still supports) the Angaité and other groups in their struggle towards the territorial restoration.

As other indigenous groups of the Chaco, the Angaité society was, to a certain extent still is, relatively egalitarian. Despite the differences among social status and the emergence of economic inequalities, men and women have equal right to be heard and participate in the local decision-making processes. This egalitarianism is not inscribed in the essence of these indigenous societies; it is an effect of the changing historical circumstances.[24] As might be deduced from Villagra's exposition on the past/traditional features of the *Enlhet-Enenlhet's* leadership, leaders were not accumulating economic benefits nor imposing their will on others. Referring to the leadership and political system of the Ayoreo, Von Bremen suggests that the traditional leader, the *dakasute,* had more of an integrative role than a representative one.[25] According to Bartolomé this contrasts with the concept of representation proposed by Pierre Clastres for "primitive societies."[26] Clastres' conception of representation contradicts the participative nature of decision-making processes in traditional societies without central political power, where the community is not delegating the representation to any individual.[27]

Blaser also explored how development and patronage networks have contributed to the emergence of new forms of leadership among the *Yshiro* people of the Paraguayan Chaco. He argues that the displacement of earlier forms of authority were provoked by the system of debt-bondage common in the area, the changes in values, family patterns and livelihood strategies as well as development and religious patronage. He states two reasons that made these new forms of leadership ill-suited to counter the centrifugal forces operating in the communities since the arrival of the whites. First, due to their position "*at the entry point of circulation networks*" their authority was very limited and highly dependent on their capacity to respond to their *patrones´* demand. Second, due to the proliferation of new leaders that found themselves a *patrón* among the many new *patrones* with new demands that appeared in the scene.[28] This redefinition of the leader's role, related to the imposition of political logic of external *patrones* (missionaries, state institutions, ranchers, NGOs), caused deep internal conflicts and constant tension among factions.[29]

The law that most significantly regulates the interaction between state and indigenous people is the Law 904/81. It fixes the socio-political organization at the level of the "indigenous communities", regulates the procedures of indigenous leader "recognition" by the INDI and establishes the "communal tenure" of land. As indicated by many scholars, a huge international indigenist movement is behind the achievement of this law in the Paraguayan context.[30] The legal representation of indigenous leaders is understood, in this law, as the result of the consensus expressed in the communal/village assembly. This means that the active and dynamic character of traditional leadership, once based on relatedness and conviviality, becomes formalized and institutionalized.[31] Two political notions were introduced to the indigenous societies: "representative leader" and "community." The second notion includes all inhabitants of a village or settlement disregarding their kinship relations or subgroups that constitute that village or settlement.[32]

Since then, indigenous leaders are considered by the state institutions as the absolute mediators for the implementation of state programs and development projects. As well documented by Bonifacio[33] and Blaser[34], EU-Paraguay bilateral funded projects promoted the foundation of indigenous organisations as mediators for the development projects. One such project is the Prodechaco, generally speaking it aimed at supporting the development of the Paraguayan Chaco since 1995. Bonifacio argues that by promoting the creation of indigenous organisations the international development cooperation is "*reinforcing specific notions of what is political and what is not.*"[35] She adds that by abiding the nature-culture divide the international development cooperation ignores a variety of indigenous political practices (like the choqueo) by labelling them as "religious" or not recognising them at all.[36] As other development projects followed the same strategy of founding indigenous organisations, in cases such as La Patria, communities ended up having more than one local indigenous organization.

But this is not a particularity of Prodechaco or some development projects. As argued by Blaser, when the respect for cultural differences was translated in the Indigenous human rights (like the ILO convention 169 and others mentioned before), they were previously rationalized in terms of neoliberal multiculturalists postulates.[37] He defines multiculturalism as a technique used to create spaces of difference within society through which "*human 'others' are incorporated within the body-politic supposedly without having to be transformed in order to fit.*"[38] As the experience of Prodechaco shows, with the creation of leaderships and organisations that "represent" the indigenous people of the Chaco in ways that would not disrupt the underlying logic of this development project, the neoliberal multiculturalist space for cultural differences is framed within a larger "market" logic that is not up for negotiation (Blaser, 2003).

5.3.1 NGO patronage

From 1995 to 2015, a considerable number of Paraguayan or foreign NGOs related to the Angaité of La Patria. Oxfam, the Italian Coopi (Cooperazione Internazionale), the local organizations Mingara, Pci Chaco, TierraViva and Grupo Sunu are just some of them. The last two, TierraViva and Grupo Sunu, have a particular approach. Villagra and Bonifacio describe this approach as a series of practices and relations based on a form of intimacy derived from a common political and cultural agenda.[39] TierraViva's work is focused on the territorial rights and SUNU's work mainly on the cultural revitalization. TierraViva supports the indigenous communities (not only La Patria) reclaiming land, this support is mainly legal, but also through the strengthening of the CLIBCH (Coordinadora de Líderes del Bajo Chaco).[40] The leaders that are part of the CLIBCH sustain a discourse based on their ethnic diversity, the affirmation of their culture, their legal rights and the denounce of the poverty condition of the communities.[41] In short, TierraViva supports the CLIBCH and sometimes mediates to put some pressure on the government to fulfil the rights of the indigenous communities – a right to emergency aid, health assistance and education as well as territorial rights. On one hand, TierraViva is an NGO conformed by non-indigenous Paraguayans that consider indigenous practices related to shamanism, such as the use of their own language and the rituals like the *choqueo,* as an intrinsic part of the indigenous land struggle. On the other, Bonifacio mentions that these practices are not considered, by the NGO, as a vital part of the political process.[42]

The relation of Grupo Sunu with La Patria can be traced back to the "language and cultural recovery program" started by a team of people hired by the Ceaduc (*Centro de Estudios Antropológicos de la Universidad Católica*). This project aimed to promote the recovery and use of the Angaité language, part of the effort was recording myths and elder's narratives.[43] Grupo Sunu was then founded by some people involved in that project. Mr. Santiago, who was involved in this project is still Grupo Sunu's main referent in La Patria. He observed that the myth's and narratives documented by the language recovery program, mostly regarding traditional life, translated in more state and NGOs development programs, trainings (*capacitaciones)* and projects in La Patria. Other elders that were part of the recovery projects also shared the perception that video/audio recorded stories were exchanged for projects.

Mr. Ramírez, unlike Mr. Santiago, is not currently involved in Grupo Sunu's work in La Patria. When I asked to interview him, he expressed his perspective in these terms:

> "*She [someone from the recovery project] did not do it well. Now I start to understand [what she did]. She won that one. I told the history of the old times. I recorded/ transmitted it. [...] I did not hear [the story], I saw it. [...] ...there was no mosquito net, no yerba mate, they wandered a lot, my grandparents, and me to. The wild animals were all tame, [...] all the animals were tame in the past. [...] ...my grandparents were all shamans. All shamans, there was nobody that was not a shaman. My grandfather, my grandmother, my father, all shamans, and I am the only one alive now. [...] She [someone from the recovery project] was happy, we told everything in Angaité language. A lot of things fit into that small audio recorder. [...] She took everything, and then she bought a pick-up for herself, she got a lot of money.*"[i]

I realised that Mr. Santiago and Mr. Ramírez shared, from different positions, the same perspective. And then I also realised that this perspective was shared by most of the Angaité. They believed that the anthropologist, like any other *patron,* was trading something. In this case, they were trading recorded myths and narrations, pictures and videos, and they thought this was just the beginning of a longer relation.[ii]

[i] *"Ha'e ndojapo poräicha. Ko'ãga katu antendéma ahávo. Oganá ha'e péva. Amombe'u mba'éicha yma guare, ha mba'eichapa la ojeiko va'ekue. [...]Che ajapo trámite idiomaitépe. [...]...ndaipóri mosquitero, ndaipóri ka'a, heta okorre va'ekue la che abuelo kuéra, ha che hendive kuéra avei, umi vícho silvestre kuéra imansopa, [...]animal imánso va'ekue yma. Che abuelo kuéra, ha'ekuéra shamán memete va'ekue. Shamán memete, ndaipóri la ndaha'éiva shamán. El finado abuelo, che abuelo abuela memete, el finado papa shamán, ha chénte la aikove ko'ãga. [...] Ha'e ovy'aiterei, idiomaite jahechauka. Pe maquina'i, grabadora, heta oike upéva. [...] Ja oguerahajepéma la otro lado. Ha upépe ha'e ikamionétama, iplátama." Mr. Ramírez, March 2016*

[ii] *The project made a book based on these recording and this book was distributed in every Angaité village/ colony.*

At that moment, in the field, I could not go further with my analysis. Months later, reading Villagra's doctoral dissertation I learned about a second and deeper aspect of this perspective of the Angaité on their relations with external *patrones.* Analysing a series of discussions, agreements and discordances among TierraViva and the CLIBCH, Villagra discovered that the indigenous leaders had a particular perspective and that the discussions were not a simple negotiation of give and take over the resources involved in a common project. In fact, on one hand, the leader's ideas of a "project" were related to the acquisition of a permanent capacity to obtain or "generate" resources from the outside. These resources could be used for multiple purposes, including those of the CLIBCH. They perceived the CLIBCH as embodied by the Directive Commission members and as such, inseparable from their respective familiar and communitarian relationships.[44] On the other hand, training (*capacitación)* was for them understood as a process of apprenticeship of distinctive knowledge (laws and accountancy), techniques (writing, computing, driving) and ways of relating with outside relationships (foreign donors). They expected that with the right "training" those capacities could be transferred, and they would become like the Paraguayan NGO employees. Villagra argues that this perspective shows common references to other processes of acquiring knowledge. According to Kidd, for instance, the paradigmatic power relation is that of the shaman with his auxiliary spirits, where "*power is to be derived from both the outside knowledge and relationship with outside beings.*"[45] Villagra adds that it is evident to us (non-Angaité) and them that the process of knowledge and capacity acquisition between a powerful outsider, such as an NGOs worker or Paraguayan politician, and an indigenous community is not equal to that of a shaman and his apprentices when dealing with spiritual beings. In fact, for us (non-indigenous Paraguayans) the leaders-NGO relationship is a physical, actual one, and the shaman-auxiliary spirit is a metaphysical one. However, Villagra argues that "*what for us is an incommensurable ontological difference, for the leaders and their fellow villagers may be a more relative and thus soluble difference between types of relationships.*"[46] The contemporary leaders search for knowledge/capacities are perceived by the Angaité as transformative, personally transferable, outside oriented and relational in nature, and this constitute a clear analogy with the shaman's quest for the agencies provided by their auxiliary spirits.[47] This means that the Angaité still perceive their leadership from a clear "relational perspective".

5.3.2 State driven development and patronage

Around 2007, the *Vice-Ministerio de Minas y Energías*, an institution from the Paraguayan state decided to support the development of La Patria colony. Through a participatory methodology, this project identified the needs of La Patria in these terms: "*Lack of incomes. Lack of economic production. Lack of transport. Lack of tools and infrastructure. Lack of communication systems. Lack of ambulance….*".[48] The project identified the failed attempts of missionaries, state and NGOs to convert indigenous people of the Chaco into sedentary farmers as paternalist approaches that lacked consideration for the indigenous culture. However, it did recognise that due to the size and qualities of their lands indigenous people "*are forced to change their way of life to convert to small peasants and ranchers.*"[49] To face this challenge, this project aimed to promote the innovation of the traditional communal organization into an "*Associative Communal Enterprise, with commercial approach and legal identity… without the abandonment of the ethnocultural behaviour and patterns of the population.*"[50] Though the project invested considerable amounts of money in the local infrastructure in order to create job opportunities (in a local *algarrobo* flour factory) and enhance the economic development, this project also failed to achieve its expected goals. Nevertheless, the ASADEC (*Asociación Angaité de Desarrollo Comunitario*), a local organization created as a result of this process, is still functioning in La Patria. According its legal statute, the mission of the ASADEC is to function as an interlocutor with the private and public institutions as well as the national and international development agencies in order to create sustainable development that respects the cultural identity of the indigenous people and to enforce the ILO Convention 169/89 ratified as Law 234/93 in Paraguay.[51] Since 2007, the ASADEC has been functioning as a parallel structure to the communal assembly of La Patria. The fact that the communal assembly has no legal identity, which means it cannot formally receive donations or loans, allowed the strategic use of the ASADEC. Every year, the ASADEC re-elects a president in a formal assembly. In 2015, I witnessed how the current president suggested two possible ways of organising the ASADEC assembly. One, was the "Paraguayan way", the second was "our way" (*ñandehaicha).* There was one main difference, the Paraguayan way implied requesting special economic resources from state institutions or NGOs. These resources would provide food and *yerba mate*, making it a special occasion. In the second way, the leaders of La Patria would simply meet and elect one among them as the new president.

Despite the relative current freedom the Angaité had managing the ASADEC, shamans or healers had not visible open participation. In fact, the idea of including shamans or 'traditional leaders' in this kind of organisations seem to be locally unexpected. They seem to reproduce what the external actors have taught, a clear distinction of the recognised leader from other local leaders like the shamans or healers. In fact, in 2015, the National Education Ministry started to enforce a new law to improve the quality of education in indigenous communities. As part of this, they asked each indigenous community to create a Communal Education Assembly. It should be constituted, among others, by someone representing "political leaders" and someone representing "religious leaders". During a special meeting among leaders and teachers from all the villages of La Patria, that took place in *La Leona* village, the participants spent hours discussing these two categories. Some indigenous teachers, those that had previously participated in the law elaboration process, tried to explain that "political leaders" referred to "recognised leaders" and "religious leaders" to shamans. Nevertheless, the participants of the meeting in *La Leona* concluded that the first term referred to people linked to political parties and the second to Anglican or Mennonite indigenous Pastors. Shamans, as showed by Bonifacio, actively participate in the political world but in a subtle and less explicit way.[52]

During my fieldwork period, I also witnessed the implementation of a development project by the regional government office for indigenous affaires, funded by the UNDP. Despite the genuine interest and the professional experience of the people involved in the implementation, the achievement of the planned goals was visibly impeded by the time schedule, economic resources and pre-established activities, among other limitations. For example, the use of standardized questionnaires of international agencies involved led to the development of a project baseline that did not reflect the living conditions of the Angaité of La Patria. In fact, some questions were not even asked because the answer was considered "evident". On the project managers' first visit to La Patria a big meeting was organized in *La Leona* village, with leaders from all the villages. The visitors openly expressed their view of La Patria: basic needs were not satisfied; the community was "*stuck because many didn't go to school*"; and nothing could be done without monetary resources. Although the project had pre-determined activities, the visitors asked the leader which were La Patria community's needs. One leader mentioned the access loss to the hunting and gathering grounds located inside the cattle ranches. Then, some leaders said that they needed wire to make fences. Two leaders added chainsaws to the list and mentioned the terrible conditions of the internal roads. Despite the fact that due to the road conditions these visitors ended up stuck in La Patria for four days, they did not interact or attempt to listen to the Angaité in order to gain a better understanding of their situation. They assumed that the situation was clear and basically confirmed what they expected, that the situation "*was the same of almost all indigenous communities*" (June 2015, field diary). Confused by the complicated names of the institutions that backed the project, leaders ended up referring to the project as "*proyecto donaciones unidas*", mixing *naciones unidas* (from United Nations) with the word *donaciones* (donations). As the project also involved an anthropologist that fulfilled the role of advisor and mediator, the leaders ended up referring to him as the "*the owner of the project" (projecto jara).* The leaders identified him as such, though they knew that the resources were coming from foreign institutions, because it was the anthropologist who was regularly visiting the community and managing these resources.

5.3.3 Political participation and political patronage

In November 2015, national elections for new municipal authorities took place. La Patria colony is located literally at the border between two municipalities. In the past years, the Angaité from La Patria have been voting in the *Pozo Colorado* district. There is a current process underway to move the official vote to *Puerto Pinasco* district. This process did not advance enough by 2015; thus, the Angaité voted in *Pozo Colorado.* If transport is a permanent problem in La Patria, the situation is worse during election periods. Consequently, people have no way to get to the Electoral Schools, situated more than 100 Kilometres away, in order to vote. This situation was manipulated by some political parties that consider the indigenous communities as vote reserves, easy to "buy" if needed.

As usual, a few weeks before the elections took place some politicians visited La Patria. Most of them tried to convince the Angaité that they were a "different" kind of candidate, not like the "normal" politicians, instead of presenting a concrete government plan. The Angaité received all these politicians and generally did not ask or suggest any concrete proposal. Based on the conversations I had about this issue, I realised that the Angaité lacked a clear understanding of how the state, in its different levels, is supposed to function. They had, instead, a very good knowledge of how the state did function and what to expect or gain from it. In this sense, the Colorado party was linked to their neighbour and local *patrón*, former senator Bader Rachid Lichi, owner of *estancia Riacho González*. Colorado party foremen arrived to La Patria guided by the caretaker of this cattle ranch. If national authorities and politicians usually only appear some time before the elections, local *patrones* were always there. Therefore, the support to a political party is mainly based on leaders' loyalty to local *patrones,* and not to a particular political party program.

The leaders, mainly in the communal assembly, mentioned the need to have a common political candidate, in November 2015. Contradicting this intended unity, each village leader negotiated with the politicians separately. Generally speaking, the leaders got money or a cow in exchange for "their people". The cow was then shared among the inhabitants of the villages. The money normally conforms to straight and clear reciprocity norms, but leaders ended up "helping" (*ayudo)* their villagers when they need it. Besides this, each individual got a previously negotiated amount of money from the politician, on that occasion it was around 50.000 G. (8,13 Euro).[iii] Most of the villages, among them were the most populated villages of La Patria, ended up supporting the Colorado party candidate (*Karoa'í*, *Las Flores*, *La Paciencia*, *La Leona*, *Urunde'y*, *Primero de Mayo*). In other cases, like *San Fernández* and *Laguna Hũ* the villagers divided, negotiated with two politicians and ended up supporting two different parties. A leader that opposed the Colorado party told me he decided to do so because "*the Colorado candidates are not giving us any 'help' when they govern"*.[iv] Only *Puente Kaigue* village supported a non-traditional party, because, according to the leader, the people behind this party were the same ones that designed the emergency aid and social programs that most benefited La Patria. The different choices among the village leaders did not reflect internal divisions or positions. I mean by this that there was no visible relation between being "culturero" or "evangelio", "mennonite" or "anglican" or aligned with any other local faction (these factions are described in the next section). Most of the villagers of *Karova Guasú* and *Seis de Marzo* decided not to vote, not because they had a particular political position, but because their leaders - two brothers - tried to negotiate a higher price per person and ended up being left aside by the politicians. Probably, the position of the leaders was related to the fact that their father, the shaman

[iii] *Exchange rate of 01.09.2016 (1€ = 6,143 G)*

[iv] *"Los colorados no dan luego ningún ayudo cuando están en el gobierno."*

that died short after, was sick and hospitalized in Asunción. This situation and the feeling that they were not receiving enough "help" from the government, politicians and other *patrones,* may have pushed them to take this position.

Buses full of people transported "their votes" to another indigenous colony, situated close to the Transchaco road. This was done a few days before the election in order to avoid the unexpected rains that would make impossible to reach La Patria. I made a short visit to *La Herencia* colony where the people supporting the Colorado party were "collected". Around 400 inhabitants of La Patria were there. The people camped in and around the school building in *La Herencia.* Long "evangelio" ceremonies took place alongside a *choqueo,* on those night. Some tensions arose between the groups but both continued their celebrations just a hundred meters away of the other.

Unexpectedly, the Colorado party lost the election. The Colorado candidate decided to keep most of the money he collected for the campaign, including the money to transport the voters, for himself. The 400 voters from La Patria were abandoned in *La Herencia.* Four days later, these people contacted the media and denounced their situation. They said it was not the first time something like this happened. Finally, perhaps due to public pressure, the Colorado candidate offered to pay 50% of the transport cost. The other 50% was paid by the leader of *La Herencia.*

5.4 Religious patronage, the re-emergence of the *choqueo* and the expansion of the modern social collective's limits

The Anglicans held an important role since the beginning of the colonization, dispossession and enclosure of the Angaité territory. As stated before, this dispossession started in 1825 during the government of Dictator Francia. It was then reinforced by President Carlos Antonio López and finally consummated after the Triple Alliance War with the *Public Land Sale Law* of 1885. Since they established the *Makxawaya* mission to the south of Angaité territory at the turn of the century the Anglicans played the controversial role of "pacifying" the indigenous people and facilitating access routes for white settlers. These main colonisation agents were the tannin companies, landowners (ranchers), Mennonites and militaries. According to Villagra, from the perspective of the Angaité, they were not dispossessed of their territory by the state that "legally" sold it, but by explorers, ranchers and soldiers.[53]

Some present leaders of *La Patria*, remembered that many indigenous families, *Angaité, Enxet* and *Sanapaná* used to go to *Makxawaya* mission every Christmas. The Anglicans offered them food, disregarding whether they were working in a cattle ranch. Looking back at that moment, some said that they didn't know that the Anglicans were receiving donations for the Angaité and only realized decades later. This comment is typical and reveals that from the Angaité perspective, the Anglicans cheated, offering as gifts things that actually belonged to the Angaité. Related to the same period, but perhaps of greater influence on the present, is the shared memory of the first big and important development project. Around 1987, the Anglicans (according to the leader's version) transferred all their facilities to the indigenous people, who were not able to manage these resources successfully. For example, Mr. Roberto remembers that "*Makxawaya [mission] used to have a small landing strip…, [and a] hospital with foreign doctors…. They transferred everything: offices, shoes and wood workshops, bakery, the cooperative… But if you give things to indigenous people without the proper follow up, without advice, everything ends up abandoned.*"[v] As part of this transfer, the families that resettled in La Patria received food, carts, a tractor (shared with *El Estribo* colony) and a considerable amount of cattle. As mentioned before, the decrease in cattle population and eventual death of all these animals is currently recalled by the Angaité to evidence their own incapacity as ranchers. However, some Angaité argue that this incapacity was limited to the "old leaders" in contrast with the "new ones" that are now "awake".

During my fieldwork, an important and well-known Angaité shaman of La Patria died. Some Paraguayans who met him, for some Angaité as well, he represented one of the last serious/powerful shamans. During a conversation about present day shamanism in La Patria, Mr. Diego and his family explained the current situation of shamanism and the changes involved:

> "*There are some shamans, but are not completely shamans… They are not so wise. Shamans are like those who go to school. Some only finish*

[v] *"Makxawaya ko oguereko peteĩ aviación… la hospital, oĩmbaite la profesional, doctor, extranjero, ingle... Oetregapá péva, ontregapaite oficina, taller zapatería, carpintería... Panadería... Cooperativa... Pero indígena oñeme'ẽramo las cosas, ndojejapói ramo la seguimiento para acompañante, upéi reho ha opyta nde rapykuere abandonado."* Mr. Roberto, November 2005

> *second grade. They do not know too many things… They don't have too much [power]… Some entered the 'university', they 'graduated', they have a lot… they are powerful. The shamans used to be very powerful… They use to live in the forest… They made horticulture gardens… They had no animals, like cows. They had plenty of wild animals. It used to be like that in the past. And they use to eat those animal's meat, without salt… My grandmother told me many things. She was a powerful shaman. She probably met my grandfather, also shaman… Everybody used to be shaman. My grandfather used to tell what he could see, far away, in another country, 'this and that will come' he used to say, and diseases arrived. Things always arrived, he did not lie. Now is different… Sometimes the powerful shamans were fasting for 7 days, today they would not make it… There are plants that are soft, they started with those in the first month, like in school. Then they knew how to handle that and they were offered a stronger one. But they had to be strong, and a 'professor' should be there, someone else should teach them. After 'studying' a bit, they could continue by themselves. But they should not share their 'guampa' or their mug, they should not drink 'tereré' with the others.'*[vi]

Mr. Diego mentions what scholars studying shamanism in the Chaco region say about the current and constant transformation of shamanic practices. He affirms that there are still shamans in La Patria. However, they are different to the powerful shamans of the past. He refers to their power as something they "have", it is something external they appropriate. He relates the powerful shamans' generative capacity to the abundance of food (wild animals and gardens) and the protective role to the protection from external and unknown threats (diseases). He states that "*everybody used to be*

[vi] *"Oĩ mimi, pero ndaha'éi la shamán iterei… Naiñarandu guasúi. Shamán ko ha'ete escuélape jaikéva rehegua. Oĩ ningo osalva segundo grado peve. Michĩ oikuaa… Sa'i oguereko… Oĩ oikéva umi Universidad, osẽma, oguerekopaitéma…imbaretéma... Yma guare la shamán imbarete…. Ka'aguyetépe oiko, yma guare… Ojapo kokue... Ndaipóri voi animal, vaka ni mba'evete. Ka'aguy pegua vícho la heta. Péicha voi yma. Yma guare ningo. Ha péva ro'okue la ho'u hikuái, ha juky'ỹ avei. Che abuela ombo'e iterei va'ekue. Péva la shamán mbarete. Oimete ohupyty ndaje che abuelo, shamán avei… ymángo shamán memete, ha'e omombe'u voi la ohecháva mombyry, otro tetã, péva oguahẽta he'i... mba'asy ou... ha oguahẽ va'erã, ndaijapúi. Ko'ãga ningo... Oĩ ningo, oĩ 7 días ndoúi va'erã mba'eve shamán imbaretéva umíva la... ha ko'ãga ndohypytýi... Oĩ pohã ikangýva, péva primer mes oñepyrũ, eskuélaicha... upe rire ojepokuaa, orahave péicha ipochy véva reuma... Pero ne aguánte va'erã, ha oi vae'rã "profesor", nembo'e jey va'erã otro. Upéi reestudiaimi, upépe ya ijeheguinte. Ha aparte va'erã kova rereko. Ijarroaparte. Ndotereréi va'erã péicha enterovévandi." Mr. Diego, September 2015*

shaman", a common statement among elders of La Patria when discussing traditional life before the arrival of the Paraguayans. In a kind of translation effort, he compares the shaman's training to the Paraguayan schooling system and explains that "a professor" (another shaman) should be there to advise the apprentice.

Mr. Diego also clearly related the arrival of the missionaries and the Paraguayans with the weakening of shamanism:

> "*My uncle used to tell us about his capabilities (as shaman) … My grandfather made him 'study'. But we didn't anymore, we could not, the Paraguayans arrived, they were around us, in the 'estancias', we started to live in the estancias… In (estancia) Tuparendá 'The New Tribes' [missionaries] were based. It was only recently that 'The New Tribes' missionaries left San Carlos [Angaité community]. They use to come here.*"[vii]

Nobody refers to himself/herself as shaman. Neither the "culturero" leaders that are openly defending and strengthening the Angaité culture nor the people healing with plants and songs. If asked, they would deny they are shamans. "*He is a doctor, from our culture, but he is not shaman*" was the answer I once got from Mr. Perú when I asked about his father. But many are signalled by others as shamans. Others, revealing that there is not a clear line to differentiate who is or who is not a shaman, say, "*I am not a shaman, I am natural, "naturaleté" (lit. completely natural).*"[viii] Surprisingly, this category was used to refer to unexpected people. Once, for example, a friend from *Karoa'í* village asked me to get some movies in Asuncion. He explained that he wanted to see that movie about the shaman. I was not sure to which movie he was referring to. Then, he added, "*that one of the shaman that fights against the Taliban…. Rambo*!"[ix] I was first confused and surprised by this use of the term. Then I realised that for some Angaité, the movie characters (even cartoons) are considered to be real. Therefore, if Rambo is taken seriously, he is clearly a very powerful shaman.

The *choqueo* ritual had been interrupted for at least two decades and was reactivated, not without resistance, in April 2006 in *Karoa'í* village. The peak

[vii] *"Che che tio kuéra sapy'ánte otalla ka'arukue, omombe'upa ore mba'épa la.... ivale va'ekue ha'e…. Che abuelo ombostudia. Upéi ore ya no, ndaikatuvéima, oĩma paraguayo kuéra ko jere, estancia, roikóma estanciahápe... Tupã rendángo omantene... oiko pe grupo Nuevas Tribus. San Carlosgui ramoite ohopa umíva, Nuevas Tribus... Ou jepe avei ko'ápe, orrekorré..." Mr. Diego, September 2015*

[viii] *Mr. Santiago, November 2015*

[ix] *Mr. Reino, September 2015*

achievements of this re-emergence of the *choqueo* was reached in 2008-2009. *Choqueos* were organized almost every two weeks rotating among most of the villages of La Patria, except for *Colonia 24, San Fernández and Las Flores.* Hundreds of people from La Patria and other communities participated. More than ever before, the "musicians" (*músicos*) started to reappear in every village. The "cultureros", those leaders promoting the *choqueo,* argued that the *choqueo* was keeping the young generation away of the Paraguayan parties that took place in the *estancias.* During those parties, the excess of alcohol used to cause regular incidents among Paraguayans and Angaité.

However, in February 2009, some Angaité publicly denounced the increase of Pentecostal cults in several villages. According to the denouncement, "*these cults basically promote extreme exorcism rituals which include day preaching, night fasting and the occasional beating up of demonised people.*"[54] Villagra observed that in La Patria, "shamanic like" phenomena provoked an alternation between calm periods and moments of unpredictable and explosive vitality.[55] Simultaneously, Mennonites groups were increasingly relating to different villages of La Patria.

In 1987, Von Bremen analysed the adoption of the Christian religion as an adaptive response to new living conditions faced by indigenous people of the Chaco. They did it, he suggested, to appropriate the power/forces that are not only necessary but also useful to live with the whites. Following his main argument, the persistence of a hunter-gatherer mentality, he suggested that this adoption is subjected to a reinterpretation based on their own and traditional criteria.[56]

Currently, the Pentecostals external religious groups are less present. However, the Mennonites maintain strong relations with *San Fernández, Las Flores, La Paciencia, Laguna Hũ, Paraíso, Urunde'y, Karova Guasú* and *Karoa'í.* Though the peak of the *choqueo* re-emergence is over, from May 2015 to April 2016 I counted ten *choqueos,* most of them organised by the Angaité without any external support and some of which lasted more than two days. All these *choqueos* took place in *Puente Kaigue* village. Interestingly, both main leaders of *La Leona* participated in all the *choqueos* celebrated in *Puente Kaigue.* They argue that their own village is mainly "evangelio" and that they, as leaders, were not allowed by the "evangelio" villagers to organise *choqueos* there. Anglican and Mennonites are currently collaborating to "spread the word" in La Patria. It is not always like this. When the Mennonites increased their influence, tensions emerged. According to an Anglican interviewee, the Mennonites started to baptise people who were already baptised by the Anglicans. In order to avoid more internal divisions, Anglicans and Mennonites decided to collaborate once they concluded that "*basically, our belief is the same.*"[x] The Pentecostal cult, which is mainly related to an indigenous church in Argentina, is considered by an Anglican technician, who has been working with the Anglicans since the creation of La Patria, as "*extremist... it allows anyone to preach, without any previous training.*"[xi] In order to stop the Pentecostal groups from taking over the village churches built by the Anglicans, specials contracts were prepared by the Anglican technician and signed by the villagers. The contract states that the La Patria church will be Anglican for 100 years!

The Anglicans' long-lasting attempt to combat and extinguish the Angaité rituals, which started at the beginning of the last century, continues. The following statement, made by the same Anglican technician in 2016, reveals both his lack of knowledge of and tolerance for these kind of practices as well as the extent to which the *choqueo* is considered something "more than religious". He also seems to suggest that the *choqueo* could not occur without external support:

> "*…The choqueo was done by the naturals… then, the anthropologist came, and they like to see these people, and gave them provisions, animals. Then, instead of lasting three days, it was prolonged… I think this is against their culture… The people are deteriorating too much… The choqueo is like a trance, your spirit opens so that the Devil can possess you, that is the door. They become possessed, then they leave their woman, and that is generating many problems. But we are not forbidding them to do it… It's only practiced in Puente Kaigue, because in Karoa'í they forbid [Mr. Santiago] to do it.*"[xii]

Currently both "cultureros" and "evangelio" continue to coexist. Some people, even leaders, switch from one side to the other throughout the years. Both groups refer to similar augments to criticise the other. "Evangelio"

[x] *Mr. Vicente, April 2016*

[xi] *Mr. Vicente, April 2016*

[xii] *"El choqueo era usando lo que estaba naturalmente… y después, vienen los antropólogos que le gusta ver a la gente, entonces les dan provista, animales. Entonces en vez de que sea tres días no mas, se alargó. Para mí no está bien es anticultural para ellos…Ya se empieza a degenerar demasiado mucho la gente… Y en el choqueo es un trance, entonces ese te abre tu espíritu, para que el Diablo te pueda poseer, y por ahí viene la puerta. Se empieza a endemoniar, y después se cambia de mujeres y ha causado muchos problema familia por eso. Pero nosotros no les prohibimos. Solo se practica en Puente Kaigue, porque en Karoa'í le prohibieron a Mr. Santiago." Mr. Vicente, April 2016, my emphasis*

claim both that the *choqueo* is related to Satan and that the younger generation are drunkards that refuse to work. "Culturero" say that people who take part of the Pentecostal and Mennonites cults are consuming drugs, they are mishandled by the so-called Pastors and that they do not want to work anymore. The Anglican cult is considered, by both sides, as an acceptable middle point.

Mr. Perú, leader of *Puente Kaigue,* and host of all the mentioned *choqueos,* compared the "culturero" and "evangelio" positions under these terms:

> "*We have a culture of celebration, we enjoy to celebrate. In celebrations, people from different places meet, they 'enjoy' (ovy'apa)… And there (in San Fernández) your neighbour will not allow you to even enter his plot. But 'evangelio' are supposed to be good people, they should take care of their fellow, but is not like that there… they would not even greet if they see you on the street. They would not say "good morning", "how are you", it's not like that, they are deeply confused… We, in our culture, we should love even those who 'live badly', we should host them, feed them. They don't do it.*"[xiii]

Mr. Perú clearly refers to what anthropologist would term the "convivial" aspects of the Angaité culture. He relates the *choqueo* to the reproduction of a culture of hospitality, good life and celebration, as opposed to the "evangelio" who are referred to as individualist and confused people who don't know "how to live well". Mr. Santiago, a "culturero" referent, made a surprising intervention during a mediation meeting. The latter was held in *La Paciencia* village following a local conflict which involved villagers and Paraguayans. Mr. Santiago recommended that "*this kind of cases should be taken and analysed in the church, because that is why people is going to church, to learn how to behave and stop doing bad things.*"[xiv] During the same meeting another leader stated that "mixed", Angaité-Paraguayan, couple's children "*do not respect the*

[xiii] *"Ore ore cultura farra, umíva igústo jafarrea. Upépe ojotopa paite lo vesino otro ladogua, ovy'apa hikuái... Ha amo ningo, Korralónpe ndaikatúi reho la ne vesíno nopermitiri eike hagua la lote. Ha evangelio siempre iguéno va'erã, ohechakuaa va'erã hapicha, upépe no... pe calle ne ruguatĩ nde rasa mante. Nde'i mo'ãi ndéve 'buen día', 'mba'éichapa', ndaipóri upéva, odesatinambaite umi gente. Ñande ñande reko vaivéva jahayhu entero, emoguahẽ, emongaru. Hikuái no." Mr. Perú, October 2015, my emphasis*

[xiv] *"…que el caso se encargue y se analise en la iglesia, ya que para eso la gente entra ahí, para portarse bien y dejar de hacer cosas que están mal." Mr. Santiago, November 2015*

law, do not want to hear what the leader says… they do not respect anyone, because they are not believers."[xv]

The Angaité internal division into "evangelio" or 'culturero', including the arguments used to criticise or defend each position regarding the conviviality or sociality (including the relations with the Paraguayans), mirrors a situation described by Blaser among the *Yshiro*. He writes that with the loosening of patronage networks, the *Yshiro* became more vocal about the negative effects of their "Paraguayanization". Blaser reads the emergence of "cultureros" and "costales" (Pentecostales) among the *Yshiro* as "*two relatively clearly bounded and proactive responses to these feelings*".[57] I argue, paraphrasing Blaser, that among the Angaité "cultureros" embraced "cultura" and the "evangelio" some also embraced "culto" to pull themselves "*out of a life of alienation and senselessness to be reinserted into a meaningful network of sociality.*"[58]

As aforementioned, some suggest that the "culturero" leaders are using the *choqueo* to receive benefits from external *patrones*, such as the NGOs promoting cultural revitalization. I would say that this may be so but would add that the re-emergence of the *choqueo* and the concrete effects of this re-emergence transcend that strategic use. A leader openly said that he agreed to participate in Padre Zanardini's, a member of the CEADUC,[xvi] cultural and language recovery project in order to receive benefits for the community. He knew that the father had provided the Ayoreo (indigenous group of the northern part of the Paraguayan Chaco) many benefits - including roads, schools and churches. Knowledge of these benefits led him to "follow" Padre Zanardini, he provided the latter information on the traditional food, the Angaité culture and the *choqueo*. After a couple of years of being a strong 'culturero', this leader decided to join the Mennonites because the cultural issue was not motivating him anymore and the Mennonites had new projects - including schools, churches, etc. However, as Blaser put it, "*renewed networks connecting humans and non-humans*" through the *choqueo* began to emerge.[59] In La Patria, "culturero" families and people shared food and resources among each other and not with "evangelio". Even families from other indigenous communities were inserted into these renewed networks. The following evidences that this re-emergence transcends strategic manipulation or

[xv] *"…ndorrespetái la líder, nohenduséi la líderpe… no respetan a nadie porque no son creyentes."* Mr. Oviedo, November 2015

[xvi] *Centro de Estudios Antropológicos de la Universidad Católica Nuestra Señora de la Asunción*

utilization: the reappearance of musicians, the involvement of young people in these practices, the reproduction and reinterpretation of practices and values related to the convivial sociality (that goes beyond the social) and the establishment of networks of reciprocity that extends across the human/non-human divide.[60]

5.5 Conclusion

I started this chapter warning that the changes in access to as well as use and management of natural resources took place at the same time as the transition from Anglican patronage to state/NGO patronage. Due to this, the Angaité's responses to each of these phenomena cannot be easily separated. Considering this, I described and analysed, in this chapter, some specific relations between the Angaité of La Patria and external actors or *patrones* (the Paraguayan state, NGOs and religious groups). I did so in order to understand how the relation with these external actors impacted the Angaité cosmographic practices. I argue that the new state/NGOs patronage networks impacted the Angaité cosmography by redefining what is considered to be "political" and thus, reducing the limits of the social collective. At the same time, the re-emergence of the *choqueo* renewed some meaningful networks of sociality that involved non-humans and other "not so human". As a result, the limits of the modern social collective were broadened. These two parallel processes reflect the changes suffered by the Angaité leadership practices and forms of sharing. I also argue that despite the important changes suffered by the Angaité, a particular "relational perspective" persists. The latter links worlds and relations that for us, Westerners, involve incommensurable ontological differences.

I first contextualised these relations with external actors as part of "development". I conceptualized "development" as the practice and discourse that aims to (re)create modernity, has the instrumental effect of extending the state bureaucratic power and, in the case of Paraguay, also extended all-pervasive patronage systems.[61] Following this, I introduced Von Bremen's "hunter-gatherer mentality" argument, according to which the responses that the indigenous people of the Chaco gave to the new challenges were based on the same logic that underlined the relations that these groups maintained with nature.[62]

Then, in section 5.3, I put the focus on the state and NGO patronage networks during the state driven development era that started around 1989.

I considered, supported by Bartolomé and Villagra, how the indigenous leader's role was further redefined by the representation political logic of external *patrones.* More precisely, I explored how the Law 904/81 introduced two external political notions to the indigenous societies: "representative leaders" and "indigenous community". In Blaser's terms, during this era, new *patrones* (state, NGOs, *macateros,* ranchers, loggers etc.) brought different demands and new leaders emerged to meet those demands.[63] He argues that with the passing of Law 904/81 the role of the new leaders as brokers became more complex. They not only depended on their *patrones* aid and requests but also on their "representativeness" and supporters' requests.[64] Law 904/81 catalysed the substitution of "traditional leadership", based on the extended kinship, generosity, shamanic and discursive skills, with "transactional leadership", based mainly on the ability to mediate between the community and non-indigenous entities.[65] I also described the political participation in national elections and argued that there is an apparent separation between local "political" factions ("culturero", "evangelio", etc.) and the support for external political parties.

In short, indigenous leadership has undergone, since colonization, transformations aimed at creating, strengthening, contesting or avoiding alliances with the surrounding society.[66] But how exactly is development (re)creating modernity and its institutions elsewhere? And how are these patronage networks and the leadership types they promote contributing to "*the asymmetrical entanglement between the Angaité world and the modern world*"? An example that answers these questions is Bonifacio's analysis of how the international development cooperation, by creating organisations as mediators, is shaping the contemporary indigenous political practices by operating within the limits of what Latour calls the "modern constitution" and Blaser calls the "myth of modernity".[67] According to the technicians managing Prodechaco, they aimed to strengthen the indigenous leaders and organisations role as intermediaries because the indigenous people lacked representational skills. By doing so, Bonifacio argues, Prodechaco was separating the links between indigenous leaders and the spirit world, which was once the core of leadership and shamanic practices. A modern configuration of politics, constituted only by the collective of humans, was affirmed. This affirmation implies more than a simple ontological dilemma because, despite the exclusion of "nature" from the political context, the shamans keep having contact with other-than-humans and influencing political processes in the communities; therefore, expanding the "modern" limits of the "social collective".[68]

In section 5.4, I explored the religious patronage networks, starting with the historical relations with the Anglicans and ending with the transformation of the shamanic practices and re-emergence of rituals like the *choqueo* around 2006. I also explored the contemporary tensions related both to this re-emergence and to the increasing influence of Pentecostal and Mennonite's religious groups. Paraphrasing Blaser, by complementing the system of debt bondage, the religious patronage contributed to the emergence of new forms of leadership among the Angaité that were not only compatible with, but even contributed to the asymmetrical entanglement between the Angaité world and the modern world.[69] However, exploring the effects of the re-emergence of the choqueo, I argued, that as the *Yshiro* did, the Angaité of La Patria were embracing "cultura" and "evangelio" in an attempt to be reinserted into meaningful networks of sociality. As the "culturero" network of sociality involves non-humans and "not so human" others, this network was expanding the limits of the modern social collective.

Before closing this chapter, let me explore some hints towards the following question: How are, Villagras' statements about the "shamanic perspective" of current leaders, and Bonifacio's argument about the persistence of non-modern collectives related to Blaser's critiques of Von Bremen's "hunter-gatherer mentality" argument? Von Bremen argued that the responses the indigenous people gave to the new challenges they were facing were based on the same logic that underlines the relations these groups maintain with "nature". He clarified that for these groups, *"the conceptual division [of culture and nature] is not existent…".*[70] This explained, according to Von Bremen, that indigenous people were trying to know more about the projects and the white world in order to be better suited to access its powers. Blaser argued, instead, that the shifting alliances were related to the all-pervasive patronage system, not to any particular hunter-gatherer mentality. Villagra's and Bonifacio's contribution along with my ethnographic evidence, indicate that there are some similarities between Von Bremen's and Blaser's arguments. As showed, Villagras' argument regarding the CLIBCH leader's expectations from the trainings and projects simply restates Von Bremen's argument in other terms. I argue, following Bonifacio, that the Angaité seem to be living in a world that ties the "natural world" and the "white's world" together, rather than substituting the "natural world" with the "white's world". This is evidenced by the re-emergence of the choqueo, the persistence of shamanic or shamanic like practices and the transformation of leadership practices as a result of their insertion in different patronage networks. In fact, Villagra seems to agree with this, when he says that for the Angaité their relations to the "auxiliary spirits" and the NGOs are not of incommensurable ontological difference. Despite the important changes suffered by the Angaité, a particular "relational perspective" persists, keeping together worlds and relations that for us (Westerns) involve incommensurable ontological differences.

Von Bremen's central argument relativizes Blaser critique that the first did not consider the impact that the patronage networks had on these societies. Von Bremen never denied this transformation, instead he highlighted the persistence, in these societies, of a particular perspective, which differs from the Western perspective and is not based on the nature-culture divide. In fact, this is one of the main arguments of Blaser's political ontology.

---

[1] Andrew Leake, 1998: 90
[2] Arturo Escobar, 1995: 156
[3] Wolfgang, Sachs, 2010
[4] Gustavo Esteva, 2010
[5] James Ferguson, 1997: 153
[6] Mario Blaser, 2003: 17
[7] James Ferguson, 1990
[8] Mario Blaser, 2003: 94
[9] Ibid.
[10] Volker Von Bremen, 1987
[11] Ibid., 53
[12] Ibid.
[13] Mario Blaser, 2003: 132
[14] Ibid., 91
[15] Ibid., 91
[16] Hannes Kalish, 2000: 37
[17] Ibid.
[18] Rodrigo Villagra, 2010
[19] Ibid., 205
[20] Ibid., 200
[21] Ibid., 212
[22] Ibid.
[23] Ibid.

[24] Mario Blaser, 2003
[25] Volker Von Bremen, 1994
[26] Miguel Bartolomé, 2000: 295
[27] Ibid.
[28] Mario Blaser, 2003: 95
[29] Miguel Bartolomé, 2000
[30] Rodrigo Villagra, 2014; Mario Blaser, 2003
[31] Mario Blaser, 2004a
[32] Miguel Bartolomé, 2000: 297
[33] Valentina Bonifacio, 2013a
[34] Mario Blaser, 2010
[35] Valentina Bonifacio, 2013a: 510
[36] Ibid., 510
[37] Mario Blaser, 2003
[38] Ibid., 157
[39] Valentina Bonifacio and Rodrigo Villagra, 2016: 94
[40] Ibid.
[41] Ibid.
[42] Ibid.
[43] Rodrigo Villagra, 2010
[44] Ibid.
[45] Ibid., 233
[46] Ibid., 236
[47] Ibid.
[48] VMME, 2007: 30
[49] Ibid., 27
[50] Ibid., 30
[51] ASADEC, 2007
[52] Valentina Bonifacio, 2013b
[53] Rodrigo Villagra, 2010
[54] Ibid., 218
[55] Ibid.
[56] Volker Von Bremen, 1987
[57] Mario Blaser, 2010: 195
[58] Ibid.
[59] Mario Blaser, 2010: 122
[60] Mario Blaser, 2003: 117
[61] Arturo Escobar, 1995; Ferguson, 1990; Blaser, 2003
[62] Von Bremen, 1987:18
[63] Blaser, 2003:95

[64] Ibid.
[65] Valentina Bonifacio, 2013a; Rodrigo Villagra, 2014
[66] Bonifacio 2013
[67] Latour 1991, in Bonifacio 2013:511
[68] Bonifacio 2013:514
[69] Blaser, 2003:78-79
[70] Von Bremen, 1987:13

6 Tensions, conflicts and the new *reglamento*

6.1 Introduction

In this chapter, I explore the current tensions and conflicts related to new or intensified livelihood activities (timber extraction, land leases and cattle raising) and the redefinition of the "political" and the "social collective" considered in the previous chapter. I also explore the way in which the Angaité are resisting this redefinition and trying to defend their own leadership practices and the related moral logic.

The redefinition of the "political" implied a transformation of the Angaité leadership practices as well as the consolidation of the figure of the transactional and representative leader. The introduction of the external political notion of "representative leaders" materialised through concrete practices that had, as effect, the production of places/territories. In the case of La Patria, this complex changing process implied, on one hand, the production of La Patria colony as "indigenous community". La Patria colony is not only smaller than "La Patria territory" but they are two distinct places based on two distinct cosmographies. On the other hand, given the fact that La Patria is constituted by 18 villages, the introduction of the new leadership notions implied the emergence of village territories. As for the Paraguayan state, each village of La Patria has status as "indigenous community". The leaders of each one of the villages are formally allowed to interact with the state and manage benefits directed to his/her village. These benefits are ideally shared among all the villagers, constituting a concrete sharing network that overlaps and cuts other local sharing networks that involve people living in different villages (for example the "culturero" and "evangelio" networks).

The production/emergence of village territories inside the communal land property coincided with the intensification of timber extraction, land leases and cattle raising inside La Patria colony. As we know, these activities are part of the responses of the Angaité to the process of deforestation, territorial dispossession and other changes related to the access, use and management of natural resources. As it might be expected, these two parallel processes generated a considerable number of tensions and conflicts.

In my exploration and analysis, I rely on some contributions from Blaser's political ontology, broadly presented in Chapter 1, and on the contributions of some other scholars. A good example of how Blaser applies his framework is his analysis of the tensions experienced by the *Yshiro* people related with changing conceptions of sharing and the consequent emergence of a "traditionalist" *Yshiro* faction. In his analysis of these circumstances he identifies "*a point of intersection between the dominant Cartesian moral logic and the relational moral logic.*"[1]

According to Blaser, the difference between these moral logics resides in "*how and what they 'see' and with what results.*"[2] Both produce entities and relations, but, while the relational moral logic sees entities as emerging from relations, the Cartesian moral logic sees relations as emerging from self-referential entities.[3] Blaser's moral logic does not refer to specific moralities, but to the logical procedure by which moralities are derived. To recognize "the right thing to do" the modern regime of truth involves three terms: the subject, the morality that indicates the right conduct and the object of the conduct. On one hand, according to the Cartesian moral logic, in order to know what is the right thing to do the subject must know what the external object is.[4] Cartesian moral logic takes for granted that the subject and the object are self-contained entities; they pre-exist their mutual relation. On the other hand, from a relational moral logic perspective, relations are the source of morality, morality is always contextual. This implies that specific situations where several spheres of life intersect with each other define the rightness or wrongness of relations.[5] As we will see in one concrete example below, events or behaviours shaped by one of these logics, can be, and is, easily translated into the terms of the other.

In his analysis, Blaser considers the tensions that emerged regarding the way a store was managed in an *Yshiro* community. The store, promoted by indigenist institutions aimed to provide credits that would allow the *Yshiro* to break from debt-bondage relations with Paraguayan *patrones.* The *indigenist* expected that the store would be run according to a logic of management concerned with the self-preservation of the enterprise. This implied that leaders, which acted as storekeepers, should stop giving credits to those who were not paying back. However, these leaders were susceptible to the influence of kin, who kept asking for more credits. Leaders were under pressure because the reception of more development aid was conditioned by the economic performance of the store. At a certain point, leaders started to refuse credit to their most-distant relatives which gave rise to tensions and led the affected group to complain to indigenist institutions. These tensions can be at first sight described as a relational moral logic embodied by the "affected group" being curtailed by the Cartesian moral logic promoted by the leaders that managed the cooperative. In fact, this kind of disputes has been usually described, by some indigenist, as a dispute between "traditionalist" and "acculturated" groups.

Contrasting with this interpretation, Blaser suggests an alternative view of the same event. In order to explain what he means, he mentions a common idea about sharing, held by the scholars behind the "hunter-gatherer paradigm". According to this idea, for indigenous people from the Chaco, "*at least ideally, individuals have the obligation to share with all the members of the residential group.*"[6] Opposing the conception of egalitarism as innately inscribed in indigenous societies, Blaser argues that this egalitarism is an effect of contingent and changing circumstances. Strathern pointed out that sharing networks might ideally extend indefinitely, but in practice are cut by other networks.[7] The circumstances in which sharing occurs defines who is within and who is outside a given network of sharing. Different forms of sharing emerged among the *Yshiro* as a result of the intersection between changing networks at different times. Just to mention a few important moments, gender, clan adscription, age ranks and particular relations with non-human entities delimited sharing before the missionaries arrived. With the abandonment of the initiation ritual after the missionaries came, this networks of sharing changed and the nuclear family and patronage relation became the relevant networks/actants that defined how sharing should be performed.[8] Blaser argues that a similar process of transformation of the notion of sharing took place among the *Yshiro* when the leaders "*brought the functioning of the cooperative to bear on their sharing (or not sharing) with their relatives.*"[9] The leaders that acted as storekeepers, he argues, embodying a situated morality, "*arising from the intersection between Cartesian and relational moral logics.*"[10] This situated morality emerged when the leaders tried to find a balance between two different ways of understanding "the right thing to do". Leaders were dependent on their indigenist patrones' aid, which made them very susceptible to the Cartesian moral logic implicit in their *patrones* demands. The leaders were at the same time constrained by the relational moral logic implicit in their kin's demands. From this situated morality, storekeepers ended up "*giving prevalence to relations over self-referential entities (i.e., the 'proper' functioning of the cooperative) when closer relatives were involved and the opposite when distant relatives were involved.*"[11] In short, from Blaser's perspective, it was a dispute over which moral logic should have precedence, not a dispute between factions who supposedly bore distinct defining traits. He also mentions that the *Yshiro* leaders soon adopted the labels (such as "traditionalist") used by the indigenist *patrones* to differentiate the factions that resulted from this conflict. Soon after in the new community formed by the dissidents/traditionalist, they applied the same logic, previously rejected by using "traditionalism", as a relevant criterion to determine with whom they should share their resources. However, Blaser argument that "traditionalism" should not be considered solely a clever manipulation of the leaders to get benefits from the indigenist *patrones* remains relevant. In fact, he shows that some events leading to the emergence of traditionalism transcend any leader's capacity to manipulate them.[12]

6.2 The consolidation of village territories and the new *reglamento*[i]

During a leader's meeting that took place in September 2015, I officially presented the participatory mapping proposal to the leader's assembly in *La Leona*. As soon I finished, a series of long interventions from the leaders started. Mr. Santiago, my host in *Karoa'í* village, mentioned that the map could show where the inhabitants of La Patria were conducting their livelihood activities and be used in the schools. Kids would be able to see how many hectares the colony had. He also said that my work could be considered a continuation of what Andrew Leake did 20 years before. Mr. Diego, from *Tatarẽ* village, reacted to this by arguing that the mapping was important but should be done under the condition that prints of the resulting maps would be left in the community. Arguing about why the map would be useful he said: "*…this is important, we will be able to show where we are and to show to people that comes from Asunción what we need and what we lack.*"[ii] Mr. Diego's intervention reflects that the map was considered to be a communication

[i] *Reglamento means agreement in Spanish. However, the Angaité used this Spanish word with a particular meaning, as referring to "how things should be done". Therefore, I decided to keep it in Spanish to avoid a misleading translation. A reglamento as understood by the Angaité is not necessarily a written document.*

[ii] *"… lo mitã he'i upéva ndaideprovéchoi, péva iplátarãnte ojapo, ha péva ñande voi sapy'ánte jafalla, péva iñimportánte, upéicharõ reikuaáta moõpa nde reime, cualquier karai oúva paraguaýgui ikatu ehechauka ichupe, kóva che aikotevẽ ha kóva chéve ofalta." Mr. Diego, September 2015*

tool that could be used to display their livelihood practices and to get some benefits from external institutions.

The next intervention, from a leader from *Paraíso* village, touched a crucial issue. He openly referenced a conflictive situation La Patria was going through:

> *"...for example, our village is growing, and we don't know how many hectares belong to us, only the old people know what belongs to us. [...]) What can we do about it? We, as leaders, have to talk about this and try to find a reglamento, to be able to know how many hectares are ours and, if we want, <u>to fence it. To have a map would be important</u> to be able to see what belongs to each village. Now there is no fence, and people are just extracting wood where they want to. <u>There is no fence to be respected.</u> Sometimes even Paraguayans are working here, and we receive too many complaints as leaders. Sometimes we do things wrong. [...] This is what I wanted to say."*[iii]

With his intervention, this leader was openly addressing the tensions related to the unclear boundaries around each village inside La Patria. He was also openly supporting the idea of starting a negotiation to find a *reglamento* in order to fence each village territory. The map, from his perspective, would be useful for this negotiation.

As soon as Mr. Heriberto finished his intervention, Mr. Oviedo, leader of *Laguna Hũ*, said he was against any kind of village mapping:

> *"We should map La Patria as a colony. Otherwise <u>a conflict will start</u>. It's better to leave (the limit's issue) to ourselves. If we get to a reglamento we will do it afterwards. Because they are many new villages but we have a common land title. Some people are slowly trying to fence some pieces. That's why the map should be at colony level, not village level..."*[iv]

[iii] *"...por ejemplo ko'ãga la comunidad okria ohóvo, ore roimehápe nga'u, péva ha'e oikuaa la anterior guare, ore ndoroikuaái mboy hectárea pa opertenecéta roimehápe, [...] Ha mba'éicha ikatu jajapo, upéva ikatu ñande líder háicha roñomongeta ha roguata de acuerdo, mboy hectáreapa ore mba'e ha péva upe hectárea ikatu romboty. Ha iñimportante la mapa rereko, ha iñimportante reikuaa moõ pevépa opertenece ajépa. Ndaipóri la korapy, lo mitã omba'apose hápe omba'apo, ndaipóri la korapy ha ojerrespeta, avéce la ou la gente o extraño, paraguayo umíva oike omba'apo, ha pépe la líder kuéra queja avei ou. Líder kuéra jajequivoca. [...] Che upéa la ambojoapyséva."* Mr. Heriberto, September 2015

[iv] *"La Patria háichante tonohẽ. Ndovaléi oiko ñorairõ kóva. Iporã ehejánte ore decisión. Oĩpa de acuerdo ojapo haguã upe rire. Porque heta comunidad ojeintegra jey la ipyahúva, porque la título peteinte ha lo mitã oho ha oñembolote oñembolote hese. Che ajerure la mapa toĩnte komunida háicha, La Patria háichante, péva sí..."* Mr. Oviedo, September 2015

Tensions over land among villages were not new, but as Mr. Oviedo expresses, the conflict was somehow avoided. Something quite unusual happened then. Mr. Recalde, the only member of the Indian Directive Board who is still one of the strongest and most influential leaders of La Patria, openly accused Mr. Oviedo of benefiting from the *status quo* and being manipulated by the Mennonite missionaries. In a long but meaningful intervention Mr. Recalde defended the mapping proposal using these words:

> *"…I don't like what [Mr. Oviedo] says. […] He wants everything to continue as it is. I like this mapping proposal, because it is not coming from another religion. […] We are not going to speak behind his (Mr. Oviedo's) back anymore, we have to speak it loud. […] His village is controlled by the Mennonites, [Mr. Oviedo] fell under Mennonite power… […] And that is going to make us clash again, because La Patria is not Mennonite… […] This (land) was not bought by the Mennonites. This land belongs to the Anglicans! We have to realise that. I don't read a word, but I know/understand. […] People fears because nobody wants to clash/have a bad life*[v] *with their fellow.*"[vi]

After four hours of public discussions and three previous months of informal discussions, the leaders decided to support the participatory mapping I proposed. Only one main condition was established, I should not map any village limits. The process of mapping and the maps would be only the starting point and first step of a negotiation process about the limits.

The idea of a new *reglamento* that should help to solve or minimise many of the current tensions has been around since a couple of years. In 2015, however, it was still not clear what this *reglamento* should look like and what should include. Also, it was not clear if the *reglamento* would be a written or oral document. For some leaders, this *reglamento* would allow to control the Paraguayans who lived or worked in La Patria, in particular the ones related to timber extraction. Others considered a crucial step to establish clear procedures to avoid conflicts related to land leases. Some related it to the

[v] *"Bad-life" should be understood as opposed to the ideal of "good life".*

[vi] *"…Péva [Oviedo] he'íva ndachegustái. (…) Ha'e ningo oipotánte péicha oñemaneja. Che che gusta kóva pe, propuesta (de mapeo), ha'e ndaha'éi otro religión ojapóva… Nañe'ẽ mo'ãi ikupére, hatã ja'e va'erã. (…) Ha'e upépe oñemaneja, menno upépe oñecaramãva ipópe ha'e ho'a. Ha péicha jajochoka jey, a ndaha'éi menno…. (…) Ko'a (yvy) ndaha'éi menno ojogua va'ekue. Kóva anglicana mba'e! Pévana ñañeme'ẽ en cuenta... Che ndaleéi ete, ha che aikuaa. Lo mitã okyhyje porque ndoiko vaiséi lo mitã oñondive"* Mr. Recalde, *September 2015*

process of enclosure and access loss to hunt in cattle ranchers. Only one thing was clear, the village limits should be negotiated and established, and this constituted a central aspect. However, the leaders knew that this would not be easy and thus postponed the *reglamento* negotiation. This is why the communal assembly requested an exclusion of the village limits from the participatory mapping. The assembly considered that the map making process would contribute to the future negotiation process.

Almost all the leaders expressed their willingness to start a negotiation process to delimit the village territories inside the colony. The reason to do it was different. On one hand, some leaders took for granted that the original land distribution, made during the first years of settlement in La Patria, was still valid. From this perspective, the negotiation was needed in order to define and recognize the situation of the "new villages", created after the land was distributed among the "original villages". This perspective was shared mainly by leaders from those original villages. On the other hand, some leaders, mainly from the "new villages", saw the negotiation process as an opportunity to give official recognition to what they consider *de facto* village limits. From this perspective, a negotiation occurred every time a new village formed. The new negotiation was a requirement because it clarified the new village's exact limits in order to avoid future tensions and conflict.

According to my observations, a new notion of land property was consolidating in La Patria. During many private conversations and public statements, the Angaité leaders argued about the importance of respecting each other's "land property", asking permission not only to extract timber or build a paddock, but even also to go through, hunt, gather or fish: *"… it is important to make the fences, (…) and to communicate, 'I would like to enter, to work there, I would like to hunt, gather or fish there', to let us know, 'So-and-so entered there, did you give him permission?'"* [vii] All leaders saw fencing as a way to mark limits on the ground. Fences have been used historically in the Chaco to make paddocks and to avoid cattle getting lost. Fences were also used to delimit land properties. Even today, during land property limits conflicts, having constructed a fence is considered by judges as proof of possession. Therefore, a fence is the prototypical *mejora* (*improvement).* But fencing is expensive. For years the Angaité unsuccessfully tried to get fencing wire from the development projects. Only recently, some projects promoting cattle

[vii] *"...iñimportánte jajapo la korapy, (…) ha ñañocomunika, 'chéko aikese amba'apose, chéko amariskase upéva', jajoavisa, 'fulano oike upépe, ndéiko eme'ẽ permiso'" Mr. Heriberto, November 2015*

breading have included this item. In 2015, a new project offered a considerable amount of wire to each village. Suddenly, some villages started to fence territory without previous negotiation. This indicated that the new *reglamento* could not be avoided any longer

6.3 Current tensions and conflicts

6.3.1 Tensions related to land leases

The first time I met Mr. Roberto was when some leaders from the communal assembly asked me to drive them to *Colonia Ceibo*. They were invited there by local authorities who wanted to create a new commission to confront common problems that affected *Kilómetro 34* (a Paraguayan colony) and La Patria. I picked up some leaders in *Karoa'í* and *La Leona*, and afterwards went to *San Fernández*. Mr. Diego and Mr. Roberto were supposed to wait for us there. Mr. Diego jumped in and quickly found one of the last free seats at the back of the old Land Rover. He had his radio with him and was listening to someone speaking on radio *Pa´i Puku*. Mr. Roberto took his time to get ready. He stopped participating in the communal assembly a few years before and was officially not *San Fernández's* leader anymore. However, he remained an influential man, not only in his village but in many others. Some leaders commented that he took so long to get ready in order to show his authority. At that moment, Mr. Diego said that Mr. Roberto was listening to the radio because Mr. Martín, the new leader, was talking about a local conflict on air. After a few minutes Mr. Roberto joined us and found a seat at the back next to Mr. Diego. Twelve of the most active leaders of the communal assembly of La Patria were packed in there. During the 3-hour trip to *Colonia Ceibo* they happily chatted and made jokes, creating a seemingly relaxed atmosphere that concealed the controversial disputes occurring among some of them at that time.

Cattle, for the Angaité, have the potential of maximizing commensality; thus, of making the desired state of contentment and tranquillity possible. But, as stated by Villagra, this potential is ambivalent, considering that cattle can also provoke internal disputes within villages and interethnic conflicts.[13] A few years ago, some village leaders started leasing some land to *paraguayos* from *Colonia Ceibo* or even to neighbouring ranchers. Land leases for cattle have proved to be a very sensitive issue due to the series of tensions and conflicts that emerged since it started. It has to be added that land lease related conflicts have sadly become common in many indigenous communities in Paraguay.

An interesting entry point into this discussion is the land lease case involving Mr. Roberto, the strong leader of *San Fernández* village, and Félix Larrosa, a *Paraguayan* from *Colonia Ceibo*. *San Fernández* is the only village with a big and well-maintained paddock. As mentioned in Chapter 4, Mr. Roberto received an established amount of money every month for allowing the use of the paddock for around 200-300 heads of cattle. This land lease started around 2013 and finished in November 2015.

In this land lease case, the oral agreement was not submitted to the communal assembly approval. No one used this argument to delegitimise it. In fact, for about a year no one critiqued the land lease. Apparently, the criticism started because Mr. Roberto did not share the money among all the villagers. When they realised that Mr. Roberto would not change the way he was managing the benefits, these villagers complained to the communal assembly. They said that Mr. Roberto had not consulted all the villagers prior to making the lease agreement. Almost at the same time, during a long flood, some cattle were rustled. Reacting to this, Mr. Roberto, who always used to be armed, bought a big shotgun and a dark blue police uniform. As the tension increased a young villager, Mr. Martín, emerged as a potential new leader and easily got the support of a faction of the village. This faction was composed mainly by Mr. Roberto's non-kin or distant-kin. In April 2015, during an official village assembly witnessed by public officials from the INDI, Mr. Martín was elected as new leader. Mr. Roberto, surprised by this event, rejected the assembly election and said he would not recognise Mr. Martín as the new leader. Trying to legitimise his leadership, Mr. Martín called radio *Pa'i Puku* and publicly denounced that Mr. Roberto was illegally leasing land of La Patria colony. At that time, a law proposal concerning penalties for illegal leases in indigenous communities was being discussed in the National Parliament. Three months later, Mr. Roberto ended the lease agreement with the *Paraguayo*. The cattle were moved to Saría in November 2015, another indigenous community located in the main road that connects La Patria with the *Transchaco* road.

Let's consider what other leaders and villagers of La Patria said about this conflict in *San Fernández*. Mr. Ignacio, Mr. Roberto's nephew, is the leader of *Tres Quebrachos* village. He founded that village with four other families who used to live in *San Fernández*. His version of the conflict is that it started because the money was not distributed; thus, villagers ceased "trusting" Mr.

Roberto. Mr. Ignacio also mentioned some tensions experienced by people who used to hunt, gather or fish in the area rented by Mr. Roberto: *"Every time someone entered there, (Mr. Roberto) suspected. Every time an animal was lost, (Mr. Roberto) was criticising these people, and problems emerged."*[viii] A conflict is always considered undesirable but more so among relatives. Mr. Ignacio recognises that the origin of the problem was the fact that the benefits were not shared but adds that Mr. Roberto negotiated the lease arrangement alone. *"He made this deal alone, he wanted everything for himself. Then the people upraised against him. And they made the change… they looked for a candidate (a new leader)."*[ix] Mr. Ignacio also added: *"And (land leasing) is forbidden. According to the law it is forbidden to lease communal land."*[x]

This is what Mr. Santiago said:

> "*(The arrangement) was negotiated secretly. The leaders (from the communal assembly) did not agree… (…). The old leader (Mr. Roberto) was not sharing the money with his villagers (pueblo), and* <u>*there are poverty and necessities in that village*</u>*. It was not he who made that paddock. The villagers made it. But they stayed quiet. He (the Paraguayan who is renting) is not a rancher (ganadero), he is a simple individual from Colonia Ceibo. We know him.*"[xi]

Mr. Santiago adds two more arguments to the land lease discussion mentioned by Mr. Ignacio. First, the leaders from the communal assembly do not agree to rent the land. Secondly, the paddock was not made by Mr. Roberto. Others mentioned this last argument which evidences that the notion of *mejora* (improvements) has been appropriated by the Angaité. This property notion is widely used in rural Paraguay to refer to the property right over human intervention on the land - including structures, clearing, wells and gardens.[14]

[viii] *"Oike jave péicha, odesconfiapa. Desconfiántema ojapo. Cualquier okañy la animal oñemboja la crítica…" Mr. Ignacio, October 2015*

[ix] *"Ha'e jeýnte ojapo la negocio upépe, ha'e ningo oipotapa... Péicha... upépe umi gente opu'ã hese kuéra. Upépe ojapóma la cambio... oheka máva la icandidatorã." Mr. Ignacio, October 2015*

[x] *"Ha umíva iprohibidoite. A según la léipe osẽ, la komunidápe iprohibido." Mr. Ignacio, October 2015*

[xi] *"Ha upéva ningo ñemihápente ojapo. Pero la líder kuéra noĩri de acuerdo...(…). Pe líder tuja, ha'énte la plata... nome'ẽi la puéblope, ha oĩ pobreza ha necesidad upépe. Ndaha'éi voi ha'e la ojapo va'ekue la korapy. Péva pueblo ojapo va'ekue. Ha okirirĩnte hikuái. (…) Péva ndaha'éi "ganadero" sino "particu", Ceibogua. Conocido avei." Mr. Santiago, October 2015*

Others also mentioned the fact that the Paraguayan leasing the land was not a rancher. He was a *particú,* a private individual, and worse still, "*he is as poor as us.*"[xii] From this perspective, deals should only be made with ranchers, real *patrones (patronité va'era).* When we talked about this conflict, Mr. Diego mentioned that what bothered him was also the fact that the Paraguayan renting land was planning to build a brick house (*oga material*). This was considered unacceptable for two reasons. Firstly, most of the locals could not afford to build such a house. Secondly and most importantly, it would be considered a *mejora* (improvement) that could make it even more difficult to expel the Paraguayan if the community decided to: *"How would he (Mr. Roberto) pay to expel him? The Paraguayan would say 'I spend so many millions to build this nice house, now you have to pay if you want me to leave.'"*[xiii]

After several visits and shared *tereré* rounds with Mr. Roberto, he started to talk about the sensitive conflictive situation. The lease agreement was over by that moment. Mr. Roberto complained about the way Mr. Martín openly denounced him on the radio:

> *"We could have seated and talked (about the land lease). He is not respecting me. People are always complaining on the radio and everybody listens. The Paraguayan used to live here, he is from Colonia Ceibo, he is my friend. I did not bring him secretly, I let the people know. 'That guy wants to come´ (to bring his cattle). I let them know, but now it seems that they are playing dumb."*[xiv]

Mr. Roberto felt that he did nothing wrong and that those things should have been solved internally, respecting his authority as a leader. Then he explained when all the problems started:

> *"First he (the Paraguayan) was paying every month, and the money does not stay in our hands, if you distribute the money it ends easily. But I distributed it, 15,000,000 G. (2,441 Euro[xv]), he paid that for more than 200 heads of cattle, it is not too much money. If you*

[xii] *Mr. Diego, September 2015*

[xiii] *"Mba'éichapa opagáta hese osẽmba haguã ha'ekuéra? He'íta kóva tanto millones agasta ajapo haguã oga'aporã, pepagána che aha haguã ko'águi". Mr. Diego, September 2015*

[xiv] *"Ñande ikatu jaguapy ñamongeta. Ndacherrespetái. La gente siempre ojeplaguea rádiope. Oñehendupaite péva, péva ningo oiko va'ekue ko'ápe, Ceibo-gua ningo, che amigo voi. Ndaha'éi voi la kiririĩhápe amoĩnge... aavisáma voi ningo la gente. "Tal fulano ningo ho'use", ha amombe'upáma la gente ha ha'etéko oñembotavy..." Mr. Roberto Fernández, November 2015*

[xv] *Exchange rate of 01.09.2016 (1Euro = 6,143 Guaraníes)*

> *distribute that it is not enough for the others… I talked to Felix (the Paraguayan), 'Don Felix, what can we do' and asked if he could give me a cow every few months instead of money. But people did not understand, they said that I was eating (stealing) all the money.'*[xvi]

Mr. Roberto was confirmed what others had claimed: the real problem was the money. Money is described by Mr. Roberto as something that doesn't stay in our hands. Many other Angaité also refer to money as something complicated and prefer to receive material provisions or animals in exchange for labour. Besides this, Mr. Roberto also recognised that the money was not enough for everybody, "*not enough for the others*". This means, Mr. Roberto's closest relatives got more benefits than those inhabitants that are not related or distantly related to him. However, he blames "the others" of misunderstanding his new attempt to get paid with animals, instead of cash.

Mr. Roberto added three more arguments to justify granting indigenous communities the right to lease their land. First, he said that Law 904 from year 1981, the one that forbids leasing, was written during Stroessner's dictatorship. Indigenous people were exempted from paying taxes while he was in power. Indigenous people currently pay taxes like everybody else - when buying food, traveling to Asunción, etc – and thus should have the same rights as any other tax payer. Specifically, the proprietor's right to use his land as he desires. Thirdly, he states that the government is impeding land leasing in indigenous communities to disable the indigenous people from becoming ranchers (*ganaderos)*. Beside the accuracy and coherence of Mr. Roberto's argument, it is interesting to note the underlining sentiment. Mr. Roberto expresses a dissatisfaction other indigenous leaders also feel. They believe that the rules the government established regarding what they can or cannot do with their natural resources are unfair. These regulations on what they can *do* materialise the suspicion that the government controls what the indigenous people can *be*.

The second argument used by Mr. Roberto was that Vicente, the Anglican technician was not against leasing. "*I talked to [Vicente], he told me 'no, no problem, that is something that depends on you, but you should be careful to avoid that he (the*

[xvi] *"Primero ningo opaga oho, por mensualmente opaga, ha la plata ningo ndopytái ñande po, la plata errepartita opa rei jey... Pero arrepartipa, peteĩ 15.000.000 opaga, ndahetai ningo, por más de 200 cabezas. Nde rerrepartita pe 15 millones, ndohypyty mo'ãi la otro avei... Aha añe Felix Larrosa, Don Felix, mba'e jajapóta? Porque no me das una vaquilla cada tanto en vez de plata? Ha upéva la gente nomanehái voi péicha... 'no, fulano ho'upa la plata'."* Mr. Roberto Fernández, November 2015

Paraguayan) brings stolen animals and things get complicated…"[xvii] When I interviewed Mr. Vicente, he said that most of the leaders from the communal assembly clearly opposed land leases. It is interesting that Mr. Roberto used Mr. Vicente's opinion, an Anglican technician, to legitimise his actions.

The last argument used by Mr. Roberto to defend his view is directed towards the leaders who said that Mr. Roberto should stop renting the land because it was illegal: *"If we decide to respect what the law says*", he argues, *"Why are those leaders not expelling the Paraguayans who are living since years in their villages? Why are they against the Paraguayans only sometimes?"*[xviii] With this last argument Mr. Roberto reveals what I also perceived during my fieldwork, the Angaité's judgment of certain acts depended on their context. This made it difficult to understand if someone was "in favour" or "against" land leases because most of them assumed either position depending on changing circumstances.

6.3.2 Tensions related to timber extraction

I mentioned before that the timber extraction activities increased since the year 2000. Loggers, most of them *paraguayos*, arrange deals with village leaders. This was not so problematic in 1995, considering that each village had a clearly allotted area of the colony. This delimitation was conducted by the Indian Directive Board established by the Anglican missionaries. However, as time passed many new villages evolved. The increased number of chainsaws combined with the increased number of loggers resulted in tensions and misunderstandings over the past years. In this section, I present the arguments and terms used by the Angaité and their leaders to describe the current situation and tensions.

In some cases, conflicts among leaders from the same village are caused by or related to this activity. Some leaders explained that when they arrange a deal with a logger, they only cut trees in the area that corresponds to their village. If they want to cut trees in other areas they ask permission to the corresponding leader. This permission can be denied but there seems to be no clear and fixed conditions. In *San Fernández*, Mr. Roberto had a good relationship with Adriano, a Paraguayan logger who also is *macatero*. In fact, he even allowed him to temporarily use one school room as a house and store to sell his products. Mr. Roberto, in turn, got some money for "renting" a

[xvii] *"he'i chéve 'no, tranquilo, péva ningo pendehegui odepende, pero eñatendente, ani anga ogueru animal vai ja'e ichupe, marca vai ha nde complica". Mr. Roberto, November 2015*
[xviii] *Mr. Roberto, November 2015*

piece of land next to his own house, where Adriano use to pile his *quebracho* posts. Mr. Martín (the new leader) had better relations with another Paraguayan, *Gordo*. The latter established his camp next to the internal road that he used to get trees from other villages. During 2015, Gordo was the most popular logger, not only because he made strait deals and paid good prices, but also because he was always there to help if someone needed a ride to the *Transchaco road.* The differences between Mr. Roberto and Mr. Martín, the leaders, are expounded upon below but succinctly summarized by the Angaité as, a "*lack of mutual understanding*"[xix]. To refer to another village area the Angaité use to say "*the property of another leader*"[xx], or "*a third-party paddock*"[xxi]. To cut many trees in "*the property of another leader*" without asking for permission is considered rude, a sign of "*lack of respect*"[xxii]. Leaders are not supposed to act this way because if they want to keep their authority they should respect the other leaders. If they respect other leaders their authority becomes stronger or as they say, "heavier"[xxiii]. The lack of clear boundaries makes this a complicated issue.

As these few examples may show, the conflictive situations generated by timber extraction are highly complex. This led a leader to take a formal proposition during a leaders' meeting in *La Leona* village. From his perspective, the only solution was to make an internal communal *reglamento* that would clarify how a leader and a logger should proceed if they plan to make an agreement. He suggested that only leaders should be allowed to give permissions or make arrangements and if they were absent someone would be assigned as substitute. The leader should first hold a village meeting where the decision and conditions of any arrangement would be decided among all the villagers. But this leader *reglamento* was not only intended to make sure that the whole village agreed, actually it was intended to give more authority of the leaders over the loggers. In his words:

> "*we would say, 'Marcos, you will work in the village; we will work with you; this is how you will work; how much will you pay to us for the wood; how many days are you allow to work here'. All that will be considered. This will make things lighter (clearer), you will work, you*

xix "*noñontendéi hikuái*"
xx "*otro lider mba'e*"
xxi "*potrero ajeno*"
xxii "*ndaiporivéima respeto*"
xxiii "*Ipesové*"

> *will give what you have to give and then you will be able to leave. You will not stay here long, and take too many things; nobody knows how much, that brings to many problems.*"[xxiv]

Other leaders suggested that, in this internal *reglamento*, the limits between villages should be established and, if possible, fenced. If fenced, they argued, it would be easier to respect other's paddocks. I will analyse the dynamics related to this idea of *reglamento* after presenting some other current conflicts.

Box 3. Other conflicts related to timber extraction

The villagers of *Monte Kué* and *Laguna Hū,* for example, experienced a short conflict when villagers of the first one sold some *quebracho* trees from the forest located between their village and *Laguna Hū*. The leader of *Laguna Hū,* however, did not argue that the limits were not respected. He simply didn't allow the logger to use the road in order to get the trees, arguing that it was his road, and that it would be damaged if used by the logger. After many attempts to convince this leader, villagers from *Monte Kué* decided to open an alternative road avoiding passing near *Laguna Hū*. Villagers from *Monte Kué* said that relations with villagers of *Laguna Hū* were good, and that they know that the leader "*does not read*"[xxv], but everybody knows that "*the forest is from everybody*"[xxvi].

In *Carpincho* village, the leader was worried because he could not stop some timber extraction that was taking place in what he considered land of his village. He told me that he wanted a document from a "*forestry institution*" in order to show it to the loggers who use to come and stay too long in his area. He complained about this situation with the police. When he made a legal complaint the police just came to get some money form the loggers. Other leaders also have photocopied documents that they use to legitimise their authority, even in cases when the document is just a newspaper article or a draft of a law not yet approved by the national parliament. This need for external documents reveals the weakening of local rules and the lack of authority of the leaders over Paraguayan loggers.

[xxiv] *"ro'éta ndéve Marcos nde remba'apóta la komunidápe, ore rojapóta ne ndive, la péicha nde remba'apóta. Mboy tanto nde repagáta oréve la madera... Mboy díapa nde ikatu remba'apo. Entonces upéva ohupyty. Ha upéva ivevyive, péicha oñemba'apo, entonces de ejapopa, entregapa, ha ikatúma resẽ. Ndaha'éi ke oiko, nde apére, heta voi ogueraha las cosas, ndereikuaái mboýpa, upéva problema memete." Mr. Ignacio, October 2015*

[xxv] *"no lee"*

[xxvi] *"ka'aguy ningo enterovéva mba'e"*

During one visit to Mr. Ignacio, leader of *Tres Quebrachos* village, he was furious with his own villagers. The reason was simple, while he was dealing with some issues in Asunción, some villagers made a deal with loggers without his permission. As soon as he found out he walked to Santa Rosa ranch, not too far away, and called to *Pa'i Puku* radio station. He denounced the improper behaviour of his villagers on air. In a local meeting were this issue was discussed, these villagers argued that Mr. Ignacio was not a good leader and accused him to block many projects from the state and NGOs that want to operate in the village.

6.3.3 Other leadership conflicts in *San Fernández*

A deeper analysis of the conflict between Mr. Martín and Mr. Roberto reflects other aspects about the changes that were taking place not only in *San Fernández* but in other villages of La Patria. As I will try to show, this conflict was not only caused by the tensions related to timber extraction and the land lease case. Tensions related to the village communal cattle and tensions among generations also contributed to the way this conflict evolved.

San Fernández has always been disputing the role of the central village with *La Leona*. In fact, many external institutions consider *San Fernández* to be the central village of La Patria. Mr. Roberto's strong and influential leadership, as well as the accessibility to his village, contribute to this fact. *San Fernández* was one of the six villages of La Patria that had some cattle in 2015. As it was mentioned, they received the cattle from a project conducted by the INDI around 2009. *San Fernández's* herd grew to 28 animals in 2014, but then, 4 were rustled by unidentified individuals.

As leader, Mr. Roberto has the moral responsibility to provide meat for special occasions and celebrations. In 2015, for example, he decided to organise the first celebration of the "Youth day" in La Patria. As on previous occasions, he avoided slaughtering one of the communal cows by convincing the owner of Santa Rosa ranch to "donate" one to support this celebration. Mr. Roberto's personal relations and everyday care of the cattle resulted in the biggest communal herd in La Patria. He identified himself as a small rancher (*pequeño ganadero)* and reinforced this identity by using rancher clothes and language expressions. At the same time, he is a very traditional leader immersed in the social construction of the ideal good and tranquil life desired by the Angaité. His experience managing the communal cattle resonates with what many other leaders told me:

> "*Cattle is always distributed (by the projects) among all the villages. What does it mean to distribute per village? It is communal. In my experience, it is not good to manage common cattle. Communal, according to the people means it belongs to everyone. You (as leader) are not able to decide how to take care because it is established in its 'reglamento' (regulation) that it is for everyone… The villagers say 'let's slaughter this one, this is ours, let's sell this one', and you, as encargado (care taker) cannot stop them. Now I think I should not inform (to the villagers) that cattle are communal, because it is not easy. It makes us suffer a lot (ñambyasy asy eterei).*"[xxvii]

Mr. Roberto, like many other leaders, finds it difficult to be the one who has to manage something that belong to the whole village. The same applies to any communal resource - the forest, the land or even the water distribution systems. But the cattle are specially complicated and more so if the *patrón* previously established, in what Mr. Roberto calls the "reglamento", how the cattle should be managed. This "reglamento" is not a written but oral agreement that outlines the way things should be done. For example, the Angaité use the same expression to say that pregnant women should not eat some things because food has it's "reglamento". The high potential of cattle as a means to maximize commensality ends up, again, creating internal suffering and disputes. Mr. Martín's critique of the way Mr. Roberto managed this cattle project reflects these disputes. He said that Roberto did not ask the village if they wanted to use the project money to buy cattle. He added that the people were tired of listening to Mr. Roberto refer to the cattle as his possessions: "*My cattle, my cows, he said, and the village was angry, but no one said anything.*"[xxviii] A lot more could be said about the communal cattle but let me go back to the main conflict and introduce the second element I mentioned above.

This conflict between Mr. Martín and Mr. Roberto is intergenerational. Mr. Martín is in his thirties and Mr. Roberto is in his sixties. About this, Mr.

[xxvii] *"Siempre ahecha, oñentregáta 'por comunidad' tanto, tanto. Mba'épa he'ise por comunida oñentregáta? Por comunitario he'i. Che experiencia ndaiporãiha voi pe comunitario ojeporu. Comunitario ningo siempre la gente he'i entero mba'e. Nde ndikatui ne rakate'ỹ porque ya oĩma la irreglamento la entero mba'erã... La comunidad he'i 'péva ñacarneáta, péva ñane mba'e, péva ñavendéta'. Ha nde como encargado ndaikatúi rejoko. …Che apensáva ndaikatuvéima che ainformá la comunitario, porque ijutu. Ñambyasy asy eterei."* Mr. Roberto, November 2015

[xxviii] *"Che mba'e, che vaka he'i, ha la komunida ipochy, pero nde'íri mba'eve"* Mr. Martín, November 2015

Martín says: *"Many old leaders don't want to leave, don't want to give the youth the possibility to manage, and as they keep ruling they become arrogant. They should say 'I am old, I don't even know how to read, and I will trust you. I will be next to you' that would be great. They have many ideas, they have experience*"[xxix] In this way, he recognises Mr. Roberto's experience but at the same time suggests that he is an inadequate leader because he does not read. Apparently reading was not as important for leaders before as it is today, considering the increasing relations and dependence of state and other external projects.

What did people from other villages say about this conflict? From the perspective of other leader, Mr. Roberto neglected too many administrative formalities that were his responsibility as leader. Mr. Martín, started to take care of these issues. He inscribed the elderly people in the state program, registered families for *Tekoporã* (the Conditional Cash Transfer program), and inscribed people in order to get an ID card (*cédula).* According to this other leader, Mr. Martín gained power though doing administrative formalities, and when the people complained against Mr. Roberto use of the money Mr. Martín was "*ready to start to make his campaign*". This other leader argued that this conflict was not unusual and that it is in this way that younger leaders get to power when old leaders don't want to freely leave.

A technician from Grupo Sunu considered that the particularity of Mr. Roberto's case is that he lost his legitimacy in the communal assembly, not only because he rented the paddock without the assembly authorization but, because he was involved in obscure negotiations with regional politicians. These negotiations resulted in a failed attempt to build a hospital and a high school for La Patria community in *San Fernández* village. This fact gave more opportunities to a young influential individual like Mr. Martín to become the new leader.

An expression used to compare old and young leaders is to call the latter "awake" and to refer to their desire for progress. This expression is mainly used by people that are or were involved with external religious groups. The Guaraní verb *hesape'a*, is normally used to refer to the action of opening the eyes, the action used to illustrate or lift someone out of ignorance. Old Angaité men usually use that expression as opposed to *naturaleté* (literally

[xxix] *"Heta lider tuja a veces nosẽséi, nome'ẽséi la mitãrusu la posibilidad la mitãrusu omaneha haguã, ha siempre upéva omanda upépe, oñembotuicha, en vez que ha'ekuéra "bueno, che che tujáma, ndalee kuaái voi, ajerovia nde rehe. Che aĩta nde ykére" Umíva igústo. Péicha iporã. Ha'e oguerekove la idea, ha'e tanto tiempo omaneja..." Mr. Martín, November 2015*

"completely natural"). For example, an old leader from *La Paciencia* told me that when he was young, kids grew up "*naturaleté, many things were not known or understood, but now we opened our eyes*". Mr. Vicente, the Anglican technician told me that he used to travel with an indigenous man that finished his bible studies so that "*he can spread the word to waken the people, something not easy to do…the old ones have to leave and give opportunity to the young who want to progress.*"[xxx] Something that differentiates the new generation, according to Mr. Santiago, is their economic perspective on and use of money: "*We indigenous people, we don't need a lot of money, just enough to live, and maybe a little bit more for our family. But now this is changing. This 'new generation' is transforming itself and is looking from a different perspective.*"[xxxi] This expression implies that, like Mr. Vicente, some Angaité believe that there is a fundamental difference between the way the old generation and the new generation perceive reality and act accordingly.

I will conclude this section by analysing how the village legitimised Mr. Martín as their leader, how Mr. Roberto resisted this handover and how external projects made everything more complicated. Villagers from San Fernández wrote an official record to the INDI stating that they elected a new leader. On April 27th, 2015, the public official from the INDI visited *San Fernández*. The village assembly explained their reasons for replacing Mr. Roberto. The latter's substitute describes the interaction as follows, "*We explained that for a long time Mr. Roberto has been <u>making us suffer, that we were missing all the projects</u> and that I was the candidate.*"[xxxii] Mr. Roberto, taking offense, rebutted the accusations against him by refusing to resign. The INDI official responded to the former leader's refusal by stating, according to Mr. Martín, "*you were not accountable to your village, you were not transparent about the projects and that is why your village is angry with you*".[xxxiii] I asked Mr. Martín if the villagers voted to elect him. "*… no, there was no vote. The village looked for someone with the right capacities: someone who created projects without being a leader. The officials from*

[xxx] *"…divulgar un poco esta cosa para que se despierte la gente, que no es una cosa fácil…los viejos ya tienen que darle lugar a los nuevos que quieren progresar." Mr. Vicente, March 2016*

[xxxi] *"Ore indígena, ore noroikotevẽi cantidad de dinero, ore rovivi haguã, ikatu opyta ore famíliape guarã. Ha ko'ãga oñecambiave oho… he'íva lo mitã la 'nueva generación' otransformáma oho. Otro háichama omaña" Mr. Santiago, October 2015*

[xxxii] *"Hace mucho tiempo Rubén orembosufri, cada proyecto rojavy, ha che la candidata." Mr. Martín, November 2015*

[xxxiii] *Mr. Martín, November 2015*

INDI has known me for a long time and this is how I ended up elected."[xxxiv] Suddenly, the projects became a central part of the problem and the solution. This reflects that the current roles of a leader include creating and managing projects. It also implies that only literate individuals are considered potential leaders. Sometimes an external institution arrives to La Patria and presents a new development project which is only implemented in some of the villages, for a number of reasons. In those cases, the Angaité from the left aside villages respond by stating "the project missed us" (*ñanderasa*), "the project avoided us" (*orevandea)* or "we missed the project" (*rojavy).* Exactly the same expression is used to refer to a storm that occurs elsewhere (*ñanderasama amâ*).

During one of my last field visits to La Patria, a sudden rain flooded the colony. The river under the bridge from *Puente Kaigue* village was so full that the current swept the road away and no one was able to transit for months. Trying to get to *San Fernández*, I stayed in *Puente Kaigue* with the leader and his family for a couple of days. Mr. Perú, the leader, didn't like to talk about other leaders but made an exception to express his support for Mr. Martín. He explained that he supported the latter because Mr. Roberto never recognised *Puente Kaigue* village territory. According to Mr. Roberto, this village was inside of the *San Fernández* paddock. A similar situation was experienced by other "young" villages (like *Tres Quebrachos* and *Primero de Mayo).* As Mr. Perú had a considerable cattle herd he was not supposed to build a paddock without Mr. Roberto's permission. This is what he said about the conflict:

> *"(Mr. Martín and Mr. Roberto) do not understand each other… Mr. Roberto is not a leader any more. He was displaced. But now he becomes arrogant… It is like that since a while… Only they will know how to deal, how to arrange… It is not from our business, we have our own village/community. They don't understand each other, and is difficult to get projects in such conditions…*"[xxxv]

Mr. Perú, again, related the leadership with external projects. Interestingly, he highlighted that every village has political autonomy. I

[xxxiv] *"…ndaha'éi por voto, sino ohechante kuri komunida, la capacidad oguereko. Ndaha'éi líder pero ojapóva voi proyecto. Ha INDIgua che conocé ymaite guive, ha apyta upépe." Mr. Martín, November 2015*

[xxxv] *"(Martín ha Roberto) noñontendéi… Karai Rubén ningo ndaha'evéima líder. Ha'éko oñenohẽma. Pero igualnte ha'e imbaretese jeýnte… Aréma péicha oiko hikuái… Ha'ekuéra mante oikuaáta mba'éicha opytáta, ha'ekuéra ojearregláta… Ore ndotokái umíva, ore aparte roguereko ore komunida. Noñoentendéi, ha la proyecto hasy ojotoka haguã…" Mr. Perú, October 2015*

mention this because some external institutions are tempted to see La Patria colony as a single political entity. However, for the Angaité, La Patria consists of 18 independent villages that collaborate on a few important issues.

When I realised the importance given to these projects, I investigated further. After a couple of long conversations, I realized that I was being retold incidents in which Mr. Roberto benefited his close kin and left aside other villagers. Many people mentioned the conflict provoked by the distribution of the "emergency kits" during the last long flood. The SEN (Secretaría de Emergencia Nacional*)* registered, following a village census elaborated by Mr. Roberto, 78 families in *San Fernández* on that occasion. Each one of these families was supposed to get one "emergency kit" every month (in October, November and December 2014). The SEN*,* signed an agreement with COOPI, an Italian NGO. COOPI was in charge of delivering 50 "emergency kits" and the SEN, the rest. The humanitarian situation in La Patria was really bad in that period, and these two institutions delivered the kits with a difference of a couple of days. Who should decide which families would get the kit first? The decision was not easy, considering that no one knew if the rest of the kits would arrive at all. Mr. Roberto, who was still the leader, decided to distribute the kits starting with his close relatives. Villagers accused him of not distributing the kits according to the village census or of giving kits to his grandchild who was not registered in the census.

One thing was clear, the projects were catalysing tensions among different groups, defined mainly but not only by kinship. The never-ending leadership conflict between Mr. Martín and Mr. Roberto was just the visible part of the two internal factions of this village. There was an apparent incapacity to solve the differences among these leaders and groups. This unresolved crisis materialized in tensions in everyday life of the village. Every time a new project was started things grew worse. This is what Mr. Ignacio said to Mr. Roberto trying to convince him to sit down and talk to Mr. Martín: "*…You are not understanding each other… and the projects come and you 'live badly', you will always 'live badly'. You will always clash, and you will affect your people too.*"[xxxvi]

Mr. Roberto believed that a solution was just around the corner. In the past, this kind of situation had been solved by dividing the village in two, by creating a new village. However, the state bureaucracy gave so much

xxxvi "…Napeñontendéi… ha la proyecto ou ha peiko vaiete, siempre peiko vaíta. Siempre pejochokáta oñondive, ha pemoĩngo vaipáta pene gente avei." Mr. Ignacio, November 2015

importance to the leadership recognition and the village census that they failed to perceive the complicated implications of this choice. Firstly, where would they go? As La Patria colony is divided among villages, a space had to be allotted to the creation of a new village. Secondly, which of the two factions should leave? After seriously considering the possibility of leaving Mr. Martín and his group decided to stay. Mr. Roberto and his group never considered the possibility of leaving. In a last attempt to solve the conflict and as a good come-back strategy, Mr. Roberto made a last proposal: *San Fernández* village should be divided in two in order to eliminate the need for anybody to move away. The division line would be easy to draw considering that he and his relatives were already living next to each other. Only a couple of families from his group would need to move. His side would be called "*San Fernández*" and Martín's side, "*Corralón*". As mentioned before, *San Fernández* village was originally called *Corralón* and many people and institutions still use the old name to refer to it.

Mr. Martín rejected Mr. Roberto's proposal. The main reason was that his new leader recognition card, issued by the INDI, said that he was leader of "*San Fernández*-Corralón":

> "*If he wants to divide the village, he has to go somewhere else… He is not a leader any more. He wants to create more problems because he does not want to let me to be in charge of the communal cattle. He wants to divide the paddock. He wants to divide in order to get projects. Like [Mr. Ignacio], he left [San Fernández] and now he gets projects*"[xxxvii]

The conflict never ended. In April 2017, Mr. Roberto was still acting as a leader for his group and using every opportunity to disqualify Mr. Martín in front of the external institutions that, not understanding the internal conflict, still went directly to Mr. Roberto's house to coordinate details of new projects.

I analyse and comment more on these cases in the conclusion of this chapter. However, it is important to state here that these cases evidence the difficulty of recognising "the right thing to do". This lack of clarity is attributed to the introduction of external political notions, the changing networks of sharing and the intensification of new livelihood activities. The Angaité villagers and their leaders traditionally used to contextually decide

[xxxvii] *Mr. Martín, November 2015*

how to distribute resources. This relational moral logic allowed them to contextually define which network or relation should be prioritised. I argue that most of the Angaité still base their actions and decisions on such a moral logic. If leadership is embodied by the leader, it is inseparable from his respective extended family and other close relationships. However, the cases considered in these chapter show that leaders are assuming a situated morality, this means they sometimes prioritise relations and sometime prioritise entities (like the law and the "indigenous community").

Also, in the past years, when a group of people felt permanently left aside by a leader, they tried to change it. To do it, they followed the procedures established by Law 904 and sustained their arguments on how a leader should behave according to the same Law. However, as soon as the new leader of the "affected faction" was recognised by the INDI they also started taking decisions based on the context and the relations involved, instead of the representative political notion. In other words, the Angaité leaders continue to prioritise the relational moral logic that characterised the traditional leadership. However, they strategically use the Law 904 to sustain their arguments in order to solve leadership conflicts and to change leaders. By doing so, they might be consolidating the Cartesian moral logic involved in the external political notions implied in this Law. This strategy is, as it might be expected, confusing the Angaité themselves.

The conflicts considered here also reflect the coexistence of different territorial visions. Older leaders, who lived and participated in the practices that produced during decades "La Patria territory", find it difficult both to accept the access loss of surrounding properties as well as to accept the emergent village territories and the fencing process. Younger leaders, who were designated as such in the past 20 years, participated in practices that produced and produce "La Patria colony" and the "village territories". This means that the tensions and conflicts are not only related to which moral logic should be prioritised and how resources should be distributed but also to how the place/territory were the everyday interaction takes place is understood.

6.4 The new *reglamento* as an attempt to overcome confusion and conflicts

There is an ongoing internal process in La Patria that tries to encourage the surfacing of the obstacles that hinder the adequate way to use, care and live there. Paraphrasing Surrallés, I argue that in order to secure a future for themselves the Angaité gradually integrated and adapted their institutions to the imposed legal system. In this process "*many of the original meanings of indigenous territorial visions and institutions have gradually become distorted.*"[15] The new *reglamento* that is currently under negotiation, can be considered an attempt to overcome the confusion and internal conflicts, in La Patria colony, but also to regain protagonism in the managing decision that affect "La Patria territory". In this last section, I describe how some leaders organised, in 2015, the first anniversary celebration. I argue that this celebration reflects how the Angaité are trying to overcome the tensions and conflicts, not only among them, but also among all the actors included in a broadly defined "social collective". The Angaité responses and positions are not homogeneous; however, there are many important shared positions.

The conflicts considered before and the ways the Angaité are handling them reveal that the they are involved in an adaptive resistance strategy. This adaptive resistance can be understood as the active process through which the Angaité coped with the massive changes they faced since colonization. They systematically adapt by making pragmatic choices. In order to explain what I mean let me shortly expose one supposition Rubenstein developed in one of his last papers. Analysing the case of a Shuar reserve created in Ecuador in 1935, he considered how this reserve provided the basis for Shuar ethnogenesis, their ethnic identity and culture.[16] In particular, he focused on the process of, what he called, "territorialisation". Due to the Shuar population boom, the clearly defined territorial limits of the reserve and a shortage of game, they could not subsist solely on hunting and gardening any more. Rubenstein tried to understand the increasing dependence on and desire for the market experienced by the Shuar. He suggested that there was another mechanism at work, besides territorialisation and the population grow, that would provide a better understanding of the Shuar adaptation. While searching for this other mechanism Rubenstein realised that most Shuar explained their historic territorial dispossession by saying that their father or grandfather had traded the land away.[17] His close friend and informant once said: "*Now, if I went there, they would not even give me a coca-cola!*"[18] Trying to understand his friend's expression, Rubenstein stated:

> "*As with all Shuar [...] he understood that when a Shuar exchanged something in return for land, he believed that he was entering into a social relationship based on the periodic exchange of gifts. Whereas the Shuar believed that the exchange signalled the beginning of a*

> *relationship, for the settlers it signalled the end of a relationship, that is, an act that, no matter how equitable, was fundamentally antisocial...*"[19]

Rubenstein suggests that the Shuar's only defence against this antisocial settler mentality was to claim an inalienable title to their own land and create a reserve. However, through this adaptive resistance (to defend themselves against the territorial dispossession), the Shuar "*internalised the settler (or capitalist) cosmography, including its ontology of selfishness.*"[20] Finally, Rubenstein adds that once the Shuar made this mental leap "*they thought there was no turning back.*"[21] Rubenstein's suggestions are perfectly applicable to the Angaité case. The Angaité argue that current conflicts are related to the new generation's selfishness as well as their attempts to divide and fence La Patria colony's land property. This evidences that they are conscious of how the creation of La Patria and the introduction of external political notions are at the root of the problems they are currently experiencing. There is a striking resemblance between the way the Shuar explain their active but involuntary participation in the sale of their land and the manner the same situation is narrated in an Angaité myth. Analysing this myth, titled "*The two shamans and the chief of the cattle*", Villagra found that the Angaité assume an active role in their land dispossession. In this myth, the Angaité resemble the Shuar in that they fail to recognise an original exchange of food for land.[22]

But let us focus now on the anniversary celebration. During an assembly meeting that took place in *La Leona*, Mr. Santiago, Mr. Oviedo and the Angaité Anglican pastor Apis shared some information about a celebration organised by initiative of the first two. They wanted to celebrate La Patria anniversary on August 15th, 2015. Nobody asked why but they decided to do it that day. National and regional authorities would be invited, they said, and "*our friends the ranchers*" too. Mr. Oviedo said that the main objective of the celebration was to talk about the community's history, "*who bought the land? Which were the leaders that participated in the land struggle? Who were the first inhabitants? Where were the first villages?*"[xxxviii] Mr. Oviedo clearly argued that the new generations and the new leaders should know this history, "*so they will not be able any more to say 'I will do what I want, this is my property.'*"[xxxix] Mr. Santiago

[xxxviii] *"Quién compró la tierra? Quiénes fueron los líderes que lucharon? Quiénes fueron los primeros habitantes? Dónde estaban las primera aldeas?"* Mr. Oviedo, August 2015

[xxxix] *"Para que ya no puedan decir, yo hago lo qe quiero porque kóva che mba'e"* Mr. Oviedo, August 2015

also stated that the celebration would help the new generations know "*how to use and take care of this land. We will see how we will use, live…, but this is indigenous property and it has a "reglamento" … We know how we should live on this land.*"[xl] Pastor Apis, also one of the organizers, held a surprisingly contradictory perspective. He stated that "*We should not worry too much about the past, we had the luck to be able to study, now we learn the skills, the [state] institutions are here and we have 'the emergency.'*"[xli] By "the emergency" Pastor Apis, like many other Angaité, refers to the state emergency relief programs.

The celebration had clear objectives and was mainly related to the leadership conflicts over the use of natural resources explored in the previous sections. It is worth mentioning that this celebration was not suggested nor economically supported by any external actor but was a personal initiative of Mr. Santiago. However, the celebration also had a message for external actors, including the state. During the same meeting, Mr. Oviedo said that they wanted to show what had changed since they acquired a title for the land property. He added an accusation, "*the state will be ashamed because almost all that we have we got from private institutions.*"[xlii]

Considering the constant tensions between the Angaité involved with external religious groups, those locally called "evangelio" and those locally known as "culturero", I was surprised by the fact that Mr. Santiago (a culturero) invited the Anglican Pastor Apis and Mr. Oviedo (*Laguna Hū* village leader, associated with a Mennonite religious group) to organize this celebration. My surprise grew when I realised that the program suggested by Mr. Santiago included many Anglican religious acts. Months later I understood that some leaders, like Mr. Santiago, were using the Anglican past of the colony's history to exemplify how current leaders and new generations should live in and use this land.

Mr. Santiago included me in the organization of the event, to such extent that I had to change my previously planned field schedule. In Asunción, he dictated some invitation letters I transcribed and sent to the Anglican church, the INDI, the SEN, the Regional Government and some NGOs. This is a common strategy used by indigenous leaders today, they seek out someone who can copy and print their handwritten texts in order to attract the

[xl] *"Cómo usar y cómo cuidar esta tierra… Mba'éicha ñande jaiporúta, jaikóta, pero kóva indígena mba'e, ha oĩ reglamento… Ore roikuaa mba'éicha javivi va'erã ko yvy ko'ápe" Mr. Santiago, August 2015*
[xli] *Mr. Apis, August 2015*
[xlii] *Mr. Oviedo, August 2015*

attention of the institutions. In La Patria, Mr. Santiago asked me to transcribe his compiled notes on the colony's history and a thank you letter to the Anglicans, for their longstanding support.

Due to the relevance of this text for my analysis, I transcribed it in its full content here, translated from the original version in Guaraní. (See Spanish and Guaraní versions in Appendix 2 and 3)

> *"August 15th 2015. / La Patria Indigenous Community, Angaité Indigenous People. / La patria Indigenous community was populated between 1983 and 1986. The land was bought by the Paraguayan Anglican Missionaries. / La Patria was a virgin land, there was nothing here, only forest. It has 22.520 hectares. There were grasslands, forest, gallery forest, palm savannahs, "espartillares", lagoons and rivers. / In 1983 there were no roads, houses, there was no health dispensaries nor schools. / In that time, 1500 people came to La Patria. Now in 2015 we are more than 3000 inhabitants in the census. The Missionaries bought the land and gave back to the Angaité what it was already theirs, as stated by the National Constitution and the Law 904. / In that time there were 10 villages and 20 leaders. In 1992 the leaders reunite and designated Damasio Flores to represent the Angaité and to follow the process of the land title. For four years Damasio processed to get the land title, meanwhile he participated in 6 Anglican national congresses in Paraguay, Chile and Argentina. Once the title was ready, in 1995 the title was handed to the Angaité. / Damasio Flores recommended that 5 persons should sign the land title: Rubén Fernández, Lalicio Rojas, Agüero Fernández, Isacio Segura and Damasio Flores. The Anglican bishop Ellison designated Damasio Flores to take care of the land title. / The missionaries transferred the title in 1995 and advised the leaders how to live in this land. As the law states, La Patria is not white's property. Whites cannot just trespass and use it. Now this is property of the Angaité people. <u>La Patria is communal land, is not property of one leader, it has many owners</u>. / Now there exist 18 villages and 34 leaders responsible of La Patria. / We want to thank the Anglican missionaries. We want to ask them not to abandon us, we want the Anglicans to <u>help us to strengthen our culture and our religion</u>, through studies and training. We ask them to let the other Anglicans that live in England and other counties know about this message. There are many new kids who do not know the history of our struggle, they don't*

know that this land was not easy to get. The Anglicans left a 'reglamento' to avoid others to come and harm us, this land has a ´reglamento´. / We want to thank the Anglicans for the church they built in some villages. / We spend many years securing this land. Many things were brought to the first settlements: animals, cows, horses, carts, tractor. The old men had no experience how to manage that. Now, the new generation changed, now the new generation knows how to manage, how to sustain in order to have a good life. / END

Mr. Oviedo also asked me to make a "Word" document. It included some information from a text that he previously wrote with Mr. Vicente, the Anglican technician. I kindly accepted, interested to see what kind of information was included and highlighted in that version of La Patria history. The full version of the document, which I typed following Mr. Oviedo's dictation, is included in the Appendix 4. This version mentioned historically significant developments such as how the Indian Directive Board was conformed in 1977 in *Makxawaya* Mission; that the land was bought in 1982; that the settlement process was difficult, people survived eating palm hart and *ysypo* (liana); that there was no *yerba mate* nor soap; and that people survived by hunting and gathering. The document then listed who were members of the Indian Directive Board, who founded each village and who were the first leaders recognised by the Paraguayan state. Finally, the document listed the churches and schools of every village and the names of the indigenous pastors, separately listing the Anglican´s and Mennonite´s pastors. Mr. Oviedo changed the title, from "History of the community" to "History of the Angaité *Kovalhok* community"[xliii]. He asked me to detail from which ranches the first families came, copying a handwritten list he wrote in Spanish based on his conversation with some elders. He also requested that I differentiate the churches built by the Mennonites from those built by the Anglicans. Finally, he mentioned that the leaders were recognised by the state in 1989. I found it interesting that both Mr. Santiago and Mr. Oviedo wanted a typed and printed Spanish version of their own history to read it to their own people.

In deeper conversations with Mr. Oviedo and Mr. Santiago, both highlighted the aim of the celebration again. Mr. Santiago said that their intention was "*to let the new generation know that the Anglicans bought the land and*

[xliii] *Kovalhok is one of many Angaité pre-colonial groups.*

that it has a 'reglamento', that it is regulated by Law 904 and the ILO convention 169, that it is communal and that nobody can say 'this is my property'"'[xliv]. Apparently changing the subject, Mr. Oviedo proposed the idea of creating a new colony centre out of any existing village in order to build the secondary school and additional services that are meant to exist in all the villages. He also insisted on the need for an external community advisor. From his perspective, the applicant should be a lawyer and an anthropologist. The advisor should manage La Patria and advice on how to improve the functioning of interventions by state institutions and other development projects. Both Mr. Santiago and Mr. Oviedo, main organizers of the celebration, sought external support for their ideas. Mr. Santiago stressed the role of the Anglicans and the national and international laws. Mr. Oviedo assumed that only an external community advisor could manage La Patria by attracting and implementing potential benefits offered by the state and other development projects.

Mr. Santiago used his personal relationships with journalists working at radio *Pa'i Puku* to announce the celebration and make a public invitation on air. He was granted a full hour on *Mundo Indígena*, a radio show dedicated to the indigenous population of the Paraguayan Chaco. I was asked to drive a group of leaders there. I was not sure if the leaders had a common message because I heard diverse descriptions of the celebration before it took place. Despite the rather rough conduction of the radio journalist, he spent a considerable amount of time trying to understand what exactly was celebrated[xlv], the four leaders were able to speak a few minutes to the audience. Mr. Santiago's first words addressed the livelihood strategies and the territorial struggle. He said:

> *"…we are worried because our population is increasing, and in our area fewer people is hunting and gathering (marisca), there are some, but not too many. Sometimes our people want to hunt and gather where the ranches are… and that worries us…. We use to live there, we*

[xliv] *"…para que los nuevos sepan que los anglicanos compraron esta tierra y que tiene un reglamento, que se rige según la ley 904 y el convenio 169, que es "comunitaria", que nadie puede decir "kóva che mba'e".*

[xlv] *Sometimes, the leaders said that they were celebrating the colony 25th anniversary, but sometimes, they also mentioned the 32 years of the land title. Considering that the land was bought in 1983, in 2015 was the 32th anniversary. Also, considering that the land title was handed in 1995, in 2015 was the 20th anniversary. In fact, none of the numbers mentioned by the leaders made sense from my or the journalist perspective. Despite we both warned the leaders about this issue, this seemed to be of absolutely no importance for them.*

> *know where the wild animals are, where the honey is, some of us are specialist, we know where to get something fast…*"[xlvi]

He also reminded listeners that when they settled in La Patria, "*there was nothing there*" and explains that this means that *"La Patria is Angaité, completely Angaité"*. Finally, he addressed what he and the villagers expected from national authorities, *"to see what they would promise."*[xlvii] I was a bit surprised because I expected the leaders to have a clear agenda and demands. Instead, as in previous situations, they only referred to basic infrastructural needs like an improvement of internal road conditions, health services and wire to build fences. They did not believe it was contradictory to state that they are "*completely Angaité*", to describe how hunting and gathering was impeded after the ranches fenced their territory and to express the expectation that the state will make promises to them, in a single speech. They believed the message was clear: they were not able to continue living a "traditional life" anymore but in order to change they expected the state, responsible for their fate, to guarantee their territorial rights as well as grant them health services, schools, wire and roads.

One day before the celebration took place, tension arose among the organizers. Conflicts were mainly related to the fact that the event was a local initiative, this meant it lacked external support and resources. The people that gathered in *La Leona* village to prepare the place (build a stage, make seats, get firewood, etc.) complained that the celebration belonged to Mr. Santiago. They directed the following accusation at him, "*(he is) making us suffer, we are hungry*". However, most of the people were *Laguna Hū* villagers working under the instruction of Mr. Oviedo. Mr. Santiago, visibly affected by the lack of appropriation of the celebration as a communal event and worried about the possibility of not having enough food to offer to the guests, decided to go to Santa Rosa ranch to try to get a cow and to use the phone to call Mr. Vicente. In Santa Rosa, he realised two important things. Firstly, he discovered that Mr. Roberto was organising a parallel event, a

[xlvi] *"…ore py'apy hína, porque ore hetave roho ko'ãga, ajépa. Ha la ore zona sa'i lénto oho upéicha la maríska umíva, oĩ ningo pero ndahetavéima. Sapy'ánte ore gente kuéra, ohose la marískape oĩhape estancia… ha péva la ore py'apy, upépe ningo oĩve la… lo mitã oiko upépe voi, oikuaa la animal silvestre umíva moõpa oĩ, moõpa oĩ eíra, ha cada uno lo mitã oguereko iprofesión, moõpa pya'eve oconseguíta."* Mr. Santiago, August 2015

[xlvii] *"… upe tiempo ndaipóri mba'eve… La Patria ha'e Angaité, completamente Angaité… Después vamos a ver qué pueden prometernos…"* Mr. Oviedo, August 2015

children's day celebration, to which the Santa Rosa ranch owner had already donated a cow. Secondly, he learned that Mr. Vicente was not well informed about the importance given to the Anglicans in the celebration and was actually planning to attend a marriage celebration in *Colonia 24* village instead. An emergency meeting was organised in *La Leona* village to evaluate how to overcome this critical situation. Mr. Santiago decided to use his own resources to get some food and to request support from Grupo Sunu to make sure that there would be oil, salt, rice and tomato sauce. Mr. Oviedo made the most of this situation by promising to get some meat from another neighbouring ranch and praising his villagers for preparing the place. As counterpart, he suggested that the celebration would start that day (one day before it was expected) with a religious celebration organised by the Mennonite church from his village. Mr. Santiago and the local leaders from *La Leona* village could not reject his proposition though *La Leona* was considered a mainly Anglican village.

Grupo Sunu considered supporting the celebration important because it was a local initiative that aimed to solve existing conflicts and tensions in La Patria. As the celebration would end with a *choqueo* in *Puente Kaigue* village, supporting the celebration meant supporting Angaité culture, the main aim of Grupo Sunu. Mr. Vicente was surprised to discover the central role given to the Anglicans and publicly reprimanded Pastor Apis for organising a wedding the same day. Again, thanks to Mr. Santiago's, local initiative two external actors, which would ordinarily oppose each other, worked together to support the celebration. Let me explain why I say this. Organizations supporting the cultural reaffirmation of indigenous people are considered by some external (non-Angaité) people as being opposed to the civilizing aim of the missionaries. Though this may sound like an old-school position, it is confirmed by Mr. Vicente's statements regarding the *choqueo,* quoted in the previous chapter. This opposition was clearly experienced differently from Mr. Santiago perspective, at least based on what he said in his letter to the Anglicans: "*…we want you to help us to strengthen our culture and our religion…*" (See letter above).

This was the context in which the celebration finally took place. During the late afternoon of August 14th, 2015, Mennonite Angaité pastors from *Laguna Hū, Karova Guasú* and *Las Flores* (the three villages with more Mennonite support) led the first ever Mennonite-style[xlviii] religious act in *La*

[xlviii] *The Angaité distinguished the "Mennonite-style" and the "Anglican-style" cults.*

Leona. The stage, microphones and electric generator prepared for the Anglican centred celebration organised by the "culturero" leader Mr. Santiago were used for that purpose. This Mennonite-style event is locally known as "*culto con pista*" something very similar to a karaoke using popular songs with new Spanish and Guaraní sticky and repetitive religious lyrics. Among the favourites songs were "smash Satan's head" and "no religion could calm my being". The second one is a Christian Pentecostal song that tells the story of someone who tried different religious options before discovering the last and best one. Among the religious options, it mentions the Mormons, the communism of Karl Marx, and African witchcraft dances. These songs were quite popular in La Patria, mostly among the young people, who even bought MP3 players and pre-loaded pen-drives. For hours, the pastors and Angaité public danced and sang these songs. The event finished when the generator ran out of gas and the sound-system ran out of battery, one hour later.

The next morning, during our *tereré* conversation, Mr. Santiago confessed that he stopped the generator in order to put an end to the act without having to ask Mr. Oviedo to stop. I was surprised, I believed both Anglican and Mennonite-style cults were very similar in regard to the position they had against the *choqueo* and the Angaité culture in general. Just consider the blog-report of an Anglican confirmation that took place in June 2015. The report celebrates the fact that 120 young men and women from *Colonia 24* village "*publicly professed their faith and commitment to Jesus through being confirmed.... Standing up and being counted as a Christian is not easy in the Paraguayan Chaco, where life is seen through the distinct lens of the spirit world and the spiritual powers of the shaman, which are often feared yet also sought in times of crisis.*"[23] During our *tereré* conversation, Mr. Santiago, incredibly relaxed and focused, expressed his concern for the internal divisions generated by the different religious groups. He himself was once very close to the Anglicans but took some distance when he realised that without his own Angaité culture La Patria and the Angaité would never be "tranquil" again. From this position, he criticised the Mennonites and Pentecostal groups that started professing a different way of managing the church a couple of years ago. According to him, they hired badly trained local pastors that do not live what they preach.

The official celebration started at 10 AM on August 15th. No national or regional authority arrived. Only Mr. Vicente and three English visitors (who managed to start and finish the wedding very early) were there. Also, the president from the CLIBCH attended for one hour. The celebration started with an Anglican-style religious act. Pastor Apis read psalm 133:1-3, "*how good and pleasant it is when God's people live together in unity*". Then, representatives from seven different local churches prayed *en idioma* (in language). I expected the Anglican religious texts to be in Angaité but in fact they were in *Enxet*. By that time, the public consisted of approximately 350 Angaité from all the villages, a couple of *macateros* who heard about the celebration and came to sell some food. Contrary to the nature of the celebration, a few meters behind the stage, a group of loggers were trying to move a big *quebracho blanco* (*Schinopsis balansae)*, illegally bought in La Patria, to be used to fix a broken bridge in a nearby road. More people were expected but the local population split into three parallel events (the wedding, the children's day and the celebration).

After the break that followed the religious introduction, the historic part began. The CLIBCH president left before the second part started. An indigenous teacher was invited to read the Spanish versions of the documents that Mr. Santiago and Mr. Oviedo asked me to copy and print. This teacher, evidencing impressive reading and translating skills, enriched the text by making it accessible to the general public. Four elders, former members of the Indian Directive Board, were invited to share their memories of the settlement process in La Patria. Hours later, Mr. Santiago told me that the first elder, who spoke completely in Angaité language, "*told everything in a very clear way, as if he would have been using a computer… and this proves that there is no need to study in order to know.*"[xlix] A photo exposition was prepared using pictures facilitated by Mr. Vicente. People spent hours identifying their relatives in these pictures and commenting on them.

Mr. Oviedo's closing words reaffirmed what was already mentioned in the documents, he claimed that because the land was bought by the Anglicans "*it does not belong to anyone, it belongs to tupã our god.*"[l] Regarding the presence of the CLIBCH president, he added that "*the land-rights lawyers confront us with the ranchers, but God is a lawyer that defends you always and against anything*"[li]. Another local pastor also compared the rights-based work of the CLIBCH with the religious groups by saying that "*I was once part of the CLIBCH, but I did not find strength there"*. It was not the first time that leaders or pastors referred to the

[xlix] *"… contó todo con una claridad como si hubiese abierto su computadora… lo que muestra que no es cierto que hay que estudiar para saber." Mr. Santiago*

[l] *"Kóva ndaha'ẽi mavave mba'e, kóva tupã ñandejára mba'e". Mr. Oviedo, August 2015*

[li] *"Los abogados de la tierra enfrentan a los ganaderos pero Dios es un abogado que te defiende ante todo siempre".*

religious groups, the NGOs (either indigenist, environmentalist or developmentalist) or the state as distinct sources of power. This reflects the persistence of the relational perspective mentioned in the previous chapter.

6.5 Conclusion

In this chapter I explored the current tensions and conflicts related to the new or intensified livelihood activities and the redefinition of both the "political" and the "social collective". I also explored the way in which the Angaité were resisting this redefinition and trying to defend their own practices and their moral logic. I argue that these everyday contradictions, tensions and internal political conflicts are strongly related to the redefinition of the "political" and the "social collective". They are also linked to the concomitant preponderance of the Cartesian moral logic in detriment of the relational moral logic that traditionally characterised the Angaité leadership and cosmographical practices. The responses from the Angaité should be understood as part of a longer and broader process of *adaptive resistance* to sustain their own existence in their interaction with other worlds. In their attempt to defend their own practices and understand the world, the Angaité ended up reinforcing the Cartesian moral logic underlying the "development cosmography" of the cattle ranches and the Paraguayan state. The Angaité are conscious of these changes, worried about their implications and are trying to regain protagonism. With the creation of La Patria colony a series of changes took place. The creation of the colony implied gaining some freedom from the Anglican missionaries. However, they were supplanted soon after by an increasing dependence on other *patrones* and the state. This dependence materialised in concrete projects and programs that aimed to integrate the Angaité into the wider national society. Nevertheless, it also implied the involvement of the state, as an entity, as an "issue" that determines how a sharing network should be delimited. The state did so by applying Law 904 that included the recognition of representative leaders. Getting this recognition allowed leaders to get and manage benefits from state institutions. By accepting this leadership recognition and the validity of the state laws, the Angaité also accepted the implicit notion of leadership and community of those laws. These notions of leadership and community embody a Cartesian moral logic that contradicts the relational moral logic which most of the Angaité concrete practices are based on. However, this contrast only manifests in situations where power asymmetries and external actors make it difficult to identify the right thing to do. Let me explain what I mean, referring to the tensions and conflicts considered before.

The Angaité, as the *Yshiro*, experienced the emergence of different notions of sharing. The commonly named "traditional" form of sharing, was delimited by gender, clan adscription, age and particular relations with non-human entities. After the territorial dispossession, when the Angaité lived dispersed among the *estancias*, this notion started to change. With the growing influence of the missionaries and the abandonment of some rituals (like the *choqueo* and *vaingka),* the nuclear family gained some prevalence and became to some extent the relevant network that defined how sharing should be performed. Until this point, the Angaité and the *Yshiro* experiences were similar. The processual change in the notion of sharing continued with the creation of La Patria colony in 1983. During the first decade of La Patria, when the Angaité had the possibility of relying on their traditional livelihood strategies again, the "traditional" notion of sharing became strong again. Unfortunately, this strengthening of tradition did not last. Since 1995, the increasing dependence on the state and the recognition of the leaders implied the consolidation of a new networks of sharing defined by the state. This "state" notion of sharing takes for granted that leaders represent their "indigenous community" and that any benefit he gets should be distributed among all equally. I argued that the Angaité conflicts reflect the coexistence of many overlapped networks of sharing and two distinct moral logics.

How can these networks of sharing coexist? During the first twelve years of La Patria (1983-1995) the Angaité were able to define, in each concrete situation, "the right thing to do". This logic allowed them to decide what they shared and with whom they shared it. However, the burgeoning tensions led to the emergence of conflicts. However, the Angaité would refer to these conflicts as a "lack of understanding" among those involved. A continuous lack of understanding results in two separate sharing networks: the fission of a village and the creation of a new one.

This coexistence of different networks of sharing became more and more problematic as the dependence on the state and external *patrones* increased. Leaders had to justify their actions to their people and to the external actors in order to maintain their legitimacy. External actors' demands were imbued in the Cartesian moral logic. Local demands, in the relational moral logic.

However, the Angaité seem not to accept the underlying moral logic of the leadership and community notions imposed by the state. In other words, as the *Yshiro* leaders did, Angaité leaders and other individuals involved in relationships with external actors seem to assume a "situated morality" in order to find a balance between different understandings of what the right thing to do is. Though most Angaité still frame their actions and conduct on a relational moral logic, thus, prioritising relations over entities, there are increasingly reinforcing the Cartesian moral logic by recognising the state and its laws.

The conflict between Mr. Roberto and Mr. Martín is a good example. The villagers of *San Fernández* gave prevalence to the relational moral logic on which Mr. Roberto's old-style leadership was based. They did so as long as most of the villagers got access to the benefits that Mr. Roberto gained by occupying this position. That is why no one rejected Mr. Roberto's lease agreement for a long time. A series of situations in which a group of people felt left aside, mainly composed by distant relatives or non-relatives, led them to accuse the leader of making an agreement without consulting them which rendered it illegal. They reinforced the leadership and community notions implicit in the law by referring to the law and state institutions in their arguments.

In Blaser's words, they used the law *"to bear on their sharing (or not sharing)."*[24] This should not be interpreted as a clash between a traditional leader (who privileges his kin) and an acculturated leader who instead wants to benefit the whole village. The "situated morality" of the leaders is perfectly exemplified by Mr. Roberto's management of the communal cattle received from an INDI project. He complains that the cattle came with a *reglament*o, a clear procedure that establishes that the cattle belong to all. This is obvious from the perspective of the INDI officials that distributed the cattle, yet, it puts Mr. Roberto in a difficult situation. He manages these cattle as if they were his property and would like to have the freedom to decide what to do with them. Leaders from many villages are in the same situation as Mr. Roberto. The difference is that the other leaders were able to manage the local demands more efficiently; therefore, they avoided the emergence of a new leader.

Blaser's approach also sheds some light on conflicts related to the village limits and the process of fencing. The village territories or the village "land properties" increasingly become a current "issue" of delineating the reach and boundaries of set sharing networks. I argue that this new phenomenon was catalysed by the already mentioned new or intensified livelihood activities (cattle raising, timber extraction and land leases). These activities, as opposed to others, imply the commercialization/commoditization of natural resources and as a result have a higher conflictive potential regarding the use of the income generated. Which is the right way to distribute this money? Who has the right to get involve in timber or land lease deals? La Patria leaders seem to be convinced that establishing clear village limits will clarify the reach of these sharing networks. By doing so the leaders of La Patria are privileging, perhaps unconsciously, the Cartesian moral logic of the "private property". As contradictory as it may seem, the Angaité seem to defend the validity of their moral logic as they take decisions that end up mining it. The tensions and conflicts reveal the resistance strategy shared by most of the Angaité. Though they adopted the political structure determined by the state, thus, accepting the leader "recognition", most of them continue to frame their practices according to a traditional leadership model.

The anniversary celebration indicates that some Angaité leaders are conscious and worried about the confusion and tensions generated by the adaptive resistance process in which they are involved. The new *reglamento* can be understood as one response, one attempt to clarify and choose between various ways of understanding and managing both La Patria colony and "La Patria territory". However, this response is not homogenous. Most Angaité see their relationship with the Paraguayan state mainly in terms of receiving gifts and help, not in terms of rights. Regarding what to expect from development projects, the Angaité are expressing two contrasting positions. On one hand, some leaders affirm that the current leaders are the new generation that changed (as opposed to the old leaders) and are ready to manage both communal projects and communal things. On the other hand, other leaders affirm that they need the help of external community advisors in order to "capture" and implement new projects as well as to clarify "the right thing to do".

There is a constant tension between "culturero" and "evangelio". However, the distance between these two groups seems to be smaller from the Angaité's perspective. In fact, the anniversary celebration and the attempts to negotiate a new *reglamento* show that both groups have many things in common. Firstly, both consider that the new generation of leaders and the new generation of Angaité live in La Patria, a piece of land bought by the Anglicans and regulated by an original *reglamento.* This *reglamento* aimed to protect them from the harm that outsiders could cause. Secondly, that there is a current need to clarify how they should use, care and live in their territory. Thirdly, that the original *reglamento* is still partially valid. This *reglamento* establishes the following things for the use, care and life in La

Patria: La Patria is not the white's property. La Patria is communal property, and this is interpreted as property of many leaders. La Patria is regulated by the national (Law 904) and international law (ILO Convention 169). Finally, that La Patria is Angaité and that the Angaité should be able to strengthen their culture and religion.

The Angaité are clearly trying to regain protagonism in the decisions that determine their lives. They think that if their territorial rights are respected and their other demands to the state are satisfied (schools, health, roads and wire), the tensions among the widely defined "social collective" in which they are involved as inhabitants of "La Patria territory" could be overcome and they will be closer to achieve their goal of a good life.

---

[1] Mario Blaser, 2003: 102
[2] Ibid., 119
[3] Ibid., 119
[4] Ibid., 99
[5] Ibid., 109
[6] Ibid.
[7] Strathern; 1996: 523
[8] Mario Blaser, 2003: 111
[9] Ibid., 111
[10] Ibid., 107
[11] Ibid., 111
[12] Ibid., 115
[13] Rodrigo Villagra, 2010: 199
[14] Kreg Hetherington, 2009: 227
[15] Alexandre Surrallés and Pedro García Hierro 2005: 10
[16] Steven L. Rubenstein, 2004: 141
[17] Ibid., 141
[18] Ibid., 142
[19] Ibid., 142
[20] Ibid., 143
[21] Ibid., 143
[22] Rodrigo Villagra, 2010
[23] Church Mission Society, 2015
[24] Mario Blaser, 2010: 115

General conclusion

Throughout this dissertation I explored how the Angaité of La Patria responded to changes in access to, use and management of natural resources inside and outside their colony over the past 20 years. For analytical purposes, this main question was divided into three sub-questions: How did their livelihood strategies and land use patterns change over this period? How does this process of change in human-environmental relations manifest in the contradictions, tensions and internal political conflicts that mark everyday life in La Patria? How did external actors empower the Angaité to strengthen their livelihood strategies and manage their territory in accordance with their own priorities in order to preserve and develop their particular way of life? For a matter of clarity, I differentiated "La Patria colony" from "La Patria territory". The first one refers to the 22,520 hectares land property bought in 1983 for the Angaité. "La Patria territory" refers to area of approximately 209,000 hectares that the Angaité of La Patria have the right to use for their traditional livelihood activities but which overlaps with privately owned cattle ranches.

My answer to my research questions implies the following six main claims:

a) By *de facto* not recognising "La Patria territory", the Paraguayan state allowed and promoted over the past 20 years resource use practices by cattle ranchers that were incompatible with certain Angaité livelihood practices.

b) The Angaité responded to these changes by adapting their livelihood strategies and land use patterns. This implied their reduced involvement in some practices and their increased participation in others. As livelihood activities and ritual cosmographic practices contribute to the production of particular places/territories, the Paraguayan state obstructed the production of "La Patria territory" by privileging practices associated to the development cosmography of the cattle ranchers.

c) Cosmographical practices contribute to the production of a particular understanding of the world. Thus, the simultaneous adaptation of livelihood strategies and the new patronage networks with the state/NGOs impacted the Angaité cosmography, leadership and networks of sharing. They also redefined the "political" and the "social collective".

d) The redefinition of the "political" and the "social collective" manifest in everyday contradictions, tensions and internal political conflicts. These tensions also reflects the preponderance of the Cartesian moral logic that detriments the relational moral logic which traditionally characterised the Angaité cosmography.

e) The Angaité's response should be understood as part of a longer and wider process of *adaptive resistance* to sustain their own existence while interacting with other worlds. The Angaité's attempt to defend their own practices and understandings of the world ended up reinforcing the Cartesian moral logic that underlies the development cosmography of the Paraguayan state.

f) The adaptation of livelihood strategies and land use patterns, as well as the adjustment to the new state/NGOs patronage networks, undermine the Angaité's capacity to manage their livelihood strategies and their territory in accordance to their own priorities in order to preserve and develop their particular way of life. The Angaité are trying to regain protagonism as well as being conscious and worried about these changes and their implications.

Let me re-state and expand each one of these claims by synthetizing what I did in each chapter.

In **Chapter 1** I delineated a theoretical framework to understand the Angaité's responses. I engaged with the political ontology project to try to overcome the tendency to dichotomize human-environmental relations, the cosmological construction or representation of the environment, and the people's practical interaction with environmental resources in the context of subsistence activities.[1] In other words, I tried to overcome the Western ontological dualism that takes "nature" and "culture" as distinct domains.[2] Based on my ethnographic fieldwork experience, I conceive the Angaité territorial struggle as an ontological conflict that is about more than just land or resources. In Blaser's words, ontological conflicts involve different assumptions about "what exists".[3] To analyse the process of change of the Angaité human-environmental relations from a political ontology perspective did not imply rethinking how the Angaité "culture" related to "nature", rather, it implied reconsidering the assumed ontological division of the "human-environmental relations" concept itself. I conceived human-environmental relations as constituted by interwoven discursive, material, social and cultural dimensions.[4]

For descriptive and analytical purposes, I relied on the following concepts: ontology, cosmography, nature regimes, place/territory and livelihood strategies. I used the ontology concept as defined by Blaser: "*the unstated but implicit understanding of how reality is constituted*".[5] Ontologies are shaped through the practices and interactions of humans and non-humans.[6] Using this concept also implied questioning the routine usage of the "culture" concept due to the implicated notion of the world as one reality "out there" and many cultural interpretations and representations of it.[7] Culture as a means to think differences among human groups is, for Blaser, an inadequate concept because it takes for granted its own ontological status.[8]

Based on Arturo Escobar's contribution on "nature regimes", Ingold's "dwelling perspective" and on Tytelman approach to the "place practices" I argued that Little's cosmography concept could be updated in order to overcome some limitations. Only such an update makes it possible to use this concept in a third-generation Political Ecology framework. I used cosmography to refer to the changing ideas and practices that contribute to the production of a particular place or territory; therefore, a particular understanding of the world. Following Tytelman, I conceived "La Patria territory", "La Patria colony" and the cattle ranches as three *"distinctive patterned places, which are produced by the practices of heterogeneous assemblages of humans and non-humans."*[9] This framework allowed me to consider the Angaité responses as changes in cosmographical practices. I focused mainly on two types of cosmographical practices: ritual practices, like the *choqueo* and shamanism, and livelihood related practices.

Claim a) By *de facto* not recognising "La Patria territory", the Paraguayan state allowed and promoted over the past 20 years resource use practices by the cattle ranchers that were incompatible with certain Angaité livelihood practices.

In **Chapter 2** I made a general contextualization of the Angaité and La Patria colony. La Patria colony has 22,520 hectares and is one of the seven Angaité communities that currently exist in the Paraguayan Chaco. Some Angaité families resettled there around 1983, when the land was purchased by the Anglican church for that purpose. These families were previously living in small groups in areas that were sold to private companies by the Paraguayan state at the end of the 19th century. These companies exploited the quebracho forest to extract tannin and also introduced cattle ranching. Due to this long process of territorial dispossession, the Angaité that resettled in La Patria in 1983 had already experienced important cultural changes and had to adapt their "traditional" way of life. According to Andrew Leake, in 1995 hunting and gathering was still important despite the Anglican attempts to transform the Angaité from hunters to farmers. It is important to note that in 1995, most hunting, fishing and gathering activities were still conducted outside of the 22,520 hectares land property of La Patria colony. In fact, in 1995 the colony land property represented only 10% of the area they used for these activities. As the Angaité had and have formal rights to access and use the resources in the areas used for these activities, I decided to call it "La Patria territory".

As shown in **Chapter 3**, over the past two decades (1995-2015), the Angaité of La Patria faced important changes in their access to, use and management of natural resources. Among the main changes that affected "La Patria territory" are the creation and expansion of cattle ranches. Many of the areas where the Angaité hunted, fished and gathered in 1995 have been deforested by the ranchers that now own those areas. Even in cases when the forest resources were not destructed, the Angaité faced increasingly denied access to these territories and resources. Though the Angaité also have formal use rights over those areas, this *de jure* recognition of rights did not reflect the interventions of state institutions that are in charge of the approval of natural resource management plans and environmental permits of the cattle ranches. Through the analysis of the RIMA of some of the cattle ranches that overlap with or are close to "La Patria territory", I showed that the Angaité territorial rights and "La Patria territory" are not respected nor recognized by the Paraguayan state in practice. Among the most relevant intra-colony changes are the increased reliance on livelihood activities based on communal natural resources and the emergence or intensification of new livelihood activities. These changes implied that some people, families and villages had privileged access to resources and unequal power to decide how these resources should be accessed, used and managed.

I explored to which extent the natural resource use practices allowed and promoted by the Paraguayan state affected the continuity of the Angaité land use practices over the past 20 years. I found that the territory used by the Angaité outside their colony decreased 35,000 hectares (16.7%) from 1995 to 2015. In this period, some villages stopped hunting, fishing and gathering on the land owned by certain ranches and started to do so in others. Out of the 19 different ranches used in 1995, 13 were still used in 2015. Many villages were still hunting, fishing and gathering in the same ranches where they used to do so in 1995. A strategy already used in 1995 was still used in 2015: some village members undertook paid jobs in ranches to ensure access to hunting, fishing and gathering areas located there. This explains why the ranches that deforested a considerable part of their properties are among the most popular ranches used for *marisca* by the Angaité. This result should be considered carefully. Deforestation implies the destruction of the habitats of the animal and plants resources used by the Angaité. The Angaité are hunting more in the ranches that deforest because they experience access loss in other areas. The mentioned strategy is logically only going to work while the ranches allow the Angaité workers and their visiting relatives to hunt in those areas, or until those areas are also deforested.

Consequent to the state's indigenist policies and parallel to the aforementioned external and internal changes, the state and NGOs increasingly implemented development projects that aimed to improve the living conditions of the Angaité. These projects focused mainly on the consolidation of their leadership and productive capacities, as well as the creation of intra-colony job opportunities. This approach demoted both the socio-cultural strengthening of the Angaité and the defence of their right to access, use and manage their territories and resources to second place priorities.

Claim b) The Angaité responded to these changes by adapting their livelihood strategies and land use patterns. This implied their reduced involvement in some practices and their increased participation in others. As livelihood activities and ritual cosmographic practices contribute to the production of particular places/territories, the Paraguayan state obstructed the production of "La Patria territory" by privileging practices associated to the development cosmography of the cattle ranchers.

In **Chapter 3**, I also explored the Angaité and the ranchers' cosmography. I argued that for the Angaité, people, non-humans and other "not so humans" (like the Paraguayans) form part of a wider "social collective". For them, "nature" constitutes part of a complex network of social interactions in which man is no more than one actor among many. As in other "organic nature regimes", rituals and practices embedded in social relations establish this continuity and links. The *choqueo*, in which the role of shamans as singers was/is central, constituted and still constitutes an important cosmographical practice for the Angaité. This ritual and the shaman's powers were used to protect the communities and provide them material benefits. But interactions with non-human beings are not confined to rituals or shamans alone. The *marisca* (hunting, gathering and fishing), a central livelihood activity, involves a set of practices through which the spiritual and the material as well as the human and non-human are synthetized. These rituals, shamanism and the *marisca* are not static. Actually, as shown by Villagra, the choqueo underwent important transformations since the proliferation of cattle ranches and the subsequent dispossession of their territory.[10] For the Angaité, the multidimensional cosmos is inhabited by human beings, animals, plants and many other types of beings, called *apmovana* (auxiliary spirits) that live in the different levels of the cosmos.[11] Humans relate with these beings; this relationship can help the human partners to master the powers controlled by them. This is why, Tytelman's conception of place proved appropriate to understanding the Angaité cosmography. He defines it as *"an assemblage of enduring (but changing) patterns produced by the interlinked practices of people and non-human persons."*[12] As Angaité shamans' capacity to communicate with these auxiliary spirits was used to protect their communities and provide them with food, it is clear that the "political domain" was not restricted to the human social domain. This means that the "social collective" was not defined by a clear distinction between "nature" and "society" or "beings" and "things". Then, I presented the cosmography embodied in the cattle ranches as a central part of the "development" cosmography that was deployed in the Angaité area since the beginning of 1900. I argued that this cosmography experienced important changes since the beginning of 1990 and that these changes were visible in the new intensive ranching style.

I argued that the Angaité's current territorial struggle is better understood if conceived as an ontological conflict. The latter are "*conflicts that ensue as different worlds or ontologies strive to sustain their own existence as they interact and mingle with each other.*"[13] The Angaité voices conceive "La Patria territory" as a place they inhabit alongside animals, plants, other non-humans and "not so human" others, such as ranchers. The change of attitude of the ranchers and their new ranching practices modified the social relations; therefore, they changed "La Patria territory". The balanced or good social relations between the Angaité, the non-humans and the Paraguayans, expressed in the tranquil past period, is contrasted with the conflictive and difficult current relations. However, the current situation is not considered to be permanent or static and the Angaité are trying to change it.

In order to understand how the Angaité of La Patria were responding to changes in the access to, use of and management of natural resources inside and outside their colony, I explored and described, in **Chapter 4**, how their livelihood strategies changed in the past two decades. I considered the livelihood activities as part of the practices that contribute to producing places/territories and particular understandings of the world. In this case, they contribute to producing "La Patria territory". This approach is based on Ingold's "dwelling perspective", according to which, the perception of the environment and the knowledge of the world is obtained through a process of *enskilment* in the course of everyday practices.[14] I showed that the Angaité responded to the enclosure of "La Patria territory", caused by the deforestation and access restrictions, by: adapting their livelihood strategies, decreasing the importance of some activities/practices (hunting, fishing, gathering) and increasingly depending on state programs as well as on livelihood activities/practices conducted inside (cattle raising, land leases and timber extraction) and outside the colony (wage labour in cattle ranches).

Claim c) Cosmographical practices contribute to the production of a particular understanding of the world. Thus, the simultaneous adaptation of livelihood strategies and the new patronage networks with the state/NGOs impacted the Angaité cosmography, leadership and networks of sharing. They also redefined the "political" and the "social collective".

The changes in the access to, use of and management of natural resources took place at the same time as the transition from the Anglican patronage to the state/NGO patronage. Due to this fact, the Angaité's responses to each of these phenomena cannot be easily separated. In **Chapter 5**, I described and analysed specific relationships between the Angaité of La Patria and external actors or *patrones* (the Paraguayan state, NGOs and religious groups). I did so in order to understand how the increasing communication with these external actors impacted the Angaité cosmographic practices. I argued that the new state/NGOs patronage networks impacted the Angaité cosmography by redefining what is considered "political" and reducing the limits of the "social collective". I considered, supported by Bartolomé and Villagra, how the indigenous leader's role was further redefined by the representation political logic of external *patrones*. More precisely, I explored how the Law 904/81 introduced two external political notions to the indigenous societies: "representative leaders" and "indigenous community". In Blaser's terms, during this era, new *patrones* (state, NGOs, *macateros,* ranchers, loggers, etc.) brought demands and new leaders emerged in response to those demands.[15] Law 904/81 encouraged the substitution of "traditional leadership" - based on extended kinship, generosity, shamanic and discursive skills - with "transactional leadership" - based on the ability to mediate between the community and non-indigenous entities.[16] In short, the state and NGO reshaped the contemporary indigenous political practices by operating within the limits of what Latour calls the "modern constitution" and Blaser calls the "myth of modernity".[17] At the same time, the re emergence of the *choqueo* renewed some meaningful networks of sociality that involved non-humans and other "not so human" which also expanded the limits of the modern "social collective". These two parallel processes reflect the changes suffered by the Angaité leadership practices and forms of sharing.

I also argued that despite the important changes the Angaité suffered, a particular "relational perspective" persists, keeping together worlds and relations that for us (Westerners) involve incommensurable ontological differences. There are clear differences between Von Bremen's and Blaser's theoretical approaches and terminology. The evidenced persistence of a "relational perspective" that resonates with Von Bremen's central argument relativizes Blaser's critique of the latter because he failed to consider the impact that the patronage networks had on these societies. Von Bremen never denied this transformation but stressed the persistence, in these societies, of a particular perspective that was distinct from the Western and not based on the nature-culture divide. In fact, this is one of Blaser's main political ontology messages and arguments.

Claim d) The redefinition of the "political" and the "social collective" manifest in everyday contradictions, tensions and internal political conflicts. These tensions also reflects the preponderance of the Cartesian moral logic that detriments the relational moral logic which traditionally characterised the Angaité cosmography.

In **Chapter 6** I explored the current tensions and conflicts related to the new or intensified livelihood activities as well as the redefinition of the "political" and the "social collective". By accepting the leadership recognition and the validity of the state laws, the Angaité accepted the notions of leadership and community implied by those laws. Among these implications was the involvement of the state as an entity. This "issue" determines how a sharing network should be delimited. Historically, the Angaité experienced the emergence of different notion of sharing. This "state" notion of sharing, takes for granted that leaders represent their "indigenous community" and that any benefit he gets should be distributed among all equally. I argued that the Angaité conflicts reflect the coexistence of many overlapped networks of sharing and two distinct moral logics that determine "the right thing to do".

Blaser's political ontology proved to be very useful to understanding the "situated morality" as a way to deal with the coexistence of two distinct moral logics. This coexistence of different networks of sharing became more and more problematic as the dependence on the state and external *patrones* increased. Leaders, in order to keep their legitimacy, had to justify their actions not only to their people but also to the external actors. External actors' demands were imbued in the Cartesian moral logic and local demands, in the relational moral logic. However, the Angaité seem to deny the underlying moral logic of the leadership and community notions imposed by the state. In other words, Angaité leaders and other individuals involved in relations with external actors seem to assume a "situated morality" in order to find a balance between different understandings of "the right thing to do". Though most Angaité still frame their actions and conduct on a relational moral logic which prioritises relations over entities, they are increasingly reinforcing the Cartesian moral logic by recognising the state and it laws.

Blaser's approach also sheds some light on the conflicts related to the village limits and the current process of fencing. The villages' territories or the villages' "land properties" seems to increasingly become an "issue" related to set sharing networks reach and boundaries. I argued that this new phenomenon was catalysed by the aforementioned new or intensified livelihood activities (cattle raising, timber extraction and land leases). La Patria leaders seem to be convinced that establishing clear village limits will clarify the reach of these sharing networks. By doing so, the leaders of La Patria are privileging, perhaps unconsciously, the Cartesian moral logic of the "private property".

Claim e) The Angaité's response should be understood as part of a longer and wider process of *adaptive resistance* to sustain their own existence while interacting with other worlds. The Angaité's attempt to defend their own practices and understandings of the world ended up reinforcing the Cartesian moral logic that underlies the development cosmography of the Paraguayan state.

I also explored in **Chapter 6** the way in which the Angaité were resisting this redefinition of the "political" and the "social collective" as well as trying to defend their own practices and their moral logic. The Paraguayan state's *de facto* disregard for "La Patria territory" is evidenced by the fact that environmental permits were granted to neighbouring cattle ranches without consulting the Angaité. This recurring incident evidences how unprepared the state is to respect the implementation of the different territorial concepts of the indigenous groups that form an integral part of it. As mentioned, the unequal distribution of the impacts of environmental change reduced the capacity of the Angaité to respond to changes, including responses that attempt to modify the structural factors that make that unequal distribution possible. Under these conditions, the Angaité have resisted, trying to secure a future for themselves, gradually adapting their institutions to the national legal system. With this *adaptive resistance*, the Angaité tried to strategically take some advantages, simultaneously adapting, re-enacting and resisting in this ontological struggle through time. I defined this *adaptive resistance* as a process in which indigenous peoples' identities and mechanisms for social and cultural reproduction come to depend "*as much on repertoires of legitimation imposed by developing states and advocacy organizations as on their own political-symbolic resources.*"[18]

The tensions and conflicts reveal the resisting strategy of most of the Angaité. Though they adopted the political structure determined by the state, thus, accepting the leader "recognition", most of them continue to frame their practices according to a traditional leadership model based on a relational moral logic. If the Angaité were waiting for better epochs, when the Paraguayan government would fully recognise their existence and the persistence of their world, the strategy is apparently failing. The Angaité might have accepted the creation of La Patria colony in order to have a minimally secure place where to rebuild their social relations and keep their practices alive, while waiting for better conditions for a territorial restitution. This is what Surrallés and García Hierro argue. However, they added that the long-term nature of this *adaptive resistance* strategy implies the distortion of the original meanings of indigenous territorial visions and institutions. A situation that is highly confusing, even for the indigenous themselves, emerges (2005). The conflicts of La Patria reflect the coexistence of different territorial visions. Older leaders, who lived and participated in the practices that produced "La Patria territory" for decades, find it difficult to accept the access loss to surrounding properties, but also to accept the emergent village territories and the fencing process. Younger leaders, who were designated as such over the past 20 years, participated in practices that produced and produce "La Patria colony" and the "village territories". The tensions and conflicts are not only related to which moral logic should be prioritised and how resources should be distributed but also to how the place/territory where the everyday interaction takes place is understood.

With the intention of defending their right to manage their livelihood strategies and their territory in accordance to their own priorities in order to preserve and develop their particular way of life, the Angaité use a strategic discourse based on their cultural reaffirmation, their legally recognised rights and the denouncement of their poverty (Bonifacio/Villagra, 2016). Two NGOs referenced here, TierraViva and Grupo Sunu, have contributed substantially to the construction and articulation of such a discourse. It circumstantially empowered the Angaité but at the same time diminished the relative importance of their own political-symbolic resources. Due to the neoliberal multiculturalist nature of the international and national rights, as well as the rather essentialist conception of indigeneity and indigenous cultures implicit in those rights, this discourse and its related practices might have weakened and confused the Angaité; therefore, undermining their capacity to manage their livelihood strategies and territory according to their priorities.

This statement needs an immediate clarification. Since its foundation in 1994, the support of TierraViva has been crucial for the indigenous territorial struggle and translated in the highest regional standards on indigenous territorial restitution and the recovery of more than 160.000 hectares now owned by indigenous communities. In the same vein, the support Grupo Sunu gave to La Patria empowered cultural recovery local initiatives, such as the re-emergence of the *choqueo* ritual, and other concrete projects that according to the Angaité contributed to their territorial struggle and the defence of their life project in general. However, what I state here is that the Angaité strategically utilise these discourses because it is their sole means of creating political capital to be held over the ranchers and the Paraguayan state.

Claim f) The adaptation of livelihood strategies and land use patterns, as well as the adjustment to the new state/NGOs patronage networks, undermine the Angaité's capacity to manage their livelihood strategies and their territory in accordance to their own priorities in order to preserve and develop their particular way of life. The Angaité are trying to regain protagonism as well as being conscious and worried about these changes and their implications.

The persistence of some core livelihood activities - like hunting, fishing and gathering – should be considered an indicator of their priorities. It is a means of resisting the long-term attempts of external actors to "convert" the Angaité and "integrate" them into the national economy. For instance, the fact that the Angaité of La Patria mention repeatedly the impact of deforestation, their neighbours' change of attitude and the implicated access loss to these territories. These attempts to "convert" and "integrate" them prove that they are being forced to change their livelihood strategies and abandon some activities, such as the *marisca.* I am not saying that the Angaité want to go back in time to live as their ancestors did, my ethnography shows how some of them are clearly exploring new activities and horizons. To believe they want to "return" or "refuse to change" is part of a modernist conception of their struggle that considers their livelihood strategy "traditional" and the community, a remnant of the past. The Angaité's livelihood strategy and way of life are contemporary, not static and equally valid to other lifestyles. Besides being equally valid, their particular way of life and their territoriality are formally recognised by international and national legislation and should be respected.

However, the Angaité's territorial struggle is more than a battle over the natural resources located in the formally recognised territory! This is exactly the contribution of having analysed it as an ontological conflict. The ontological nature of the struggle is illustrated by the above-mentioned impacts that the changes of livelihood strategies and land use, combined with the impact of the patronage networks on which the Angaité are involved, had on their cosmography, leadership and forms of sharing. The challenge for this dissertation was to describe the ontological nature of this struggle without reinforcing some ontological assumptions of central Western academic categories. This led me to clarify that the role of the shamans and other leaders should not be considered solely "political" and that livelihood activities, such as the *marisca,* are not just "economical" activities. Both leadership and livelihood activities are in fact better understood as cosmographical practices that synthetize the human, the non-human and "not so human" in a particular way, distinct from our Western modern ontology.

The fact that the Angaité's cosmographies and leadership have been changing should not obscure the persistence of ontological differences. These changes have been continual but intensified since the colonization of their territory, more than a century ago. The relationships with the external actors considered here - the state, NGO and religious patrones - undermine the Angaité's capacities to manage their livelihood strategies and territory; these relations impose the modern configuration of politics as constituted only by the collective of humans, which excludes "nature" from the political context (Bonifacio 2013). This dissertation evidences that these changes implied the consolidation of the "transactional leaders" and the exclusion of the shamans or the rituals, such as the choqueo, conceived as pertaining to the "cultural" domain. This ontological dilemma might be considered unimportant but it is in fact crucial. On one hand, it impedes us (Westerns) from seeing and considering the persistence of the Angaité's links to non-human beings which influence political processes (Bonifacio 2013). On the other hand, with the imposition of the modern configuration of politics, only science, in this case represented by the SEAM and the technicians that conduct the environmental impact assessments are the ones authorized to talk about nature. The Angaité might "believe" what they want but science has the last word. In the same line, the unawareness manifested by most of the considered external actors involved in the persistence of the "relational perspective" mentioned above, constitutes a perfect breeding ground where deep misunderstandings can proliferate. Due to the unequal power relationship between the Angaité and the state or other external actors, these misunderstandings end up undermining the Angaité's capacity to preserve and develop their way of life and life project.

The Angaité are worried about these changes and their implications while at the same time trying to regain protagonism. The anniversary celebration indicates that some Angaité leaders are conscious and worried about the confusion and tensions generated by the adaptive resistance process in which they are involved. The new *reglamento* can be understood as one response, one attempt to clarify and choose between various ways of understanding and managing both La Patria colony and "La Patria territory". This response is not homogenous but local factions have many things in common. The Angaité are clearly trying to regain protagonism in the decisions that determine their life. They think that if their territorial rights are respected and their other demands to the state are satisfied (schools, health, roads and wire), the tensions among the widely defined "social collective" in which they are involved as inhabitants of "La Patria territory" could be overcome and they will be a bit closer to achieve their goal of a good life.

As stated in the introduction, this dissertation was fuelled by the sense of vacuum created by Blaser's critique of Paraguayan indigenist individuals and institutions, as well as by the broader ranging indigeneity critique. Blaser argued that the Paraguayan indigenist field shares a particular conception of indigenousness associated with a hunting-gathering way of life. This critique resonates with the indigeneity critique that also highlighted that indigeneity and indigenous practices have been characterized as expressly tied to particular environments and places.[19] I argue that my ethnography of the Angaité avoided an essentialist conception of indigenousness. It also avoided a static and nostalgic description of the Angaité, typically described in the bibliography as deculturized. My redefinition of the cosmography concept was crucial to achieve this. This definition is indebted to Tytelman's definition of *"practices of place"* as the practices that contribute to the production of place.[20] Place, based on Ingold's "dwelling perspective" is not defined in relation to an abstract space, *"but instead to the web of relations and practices that conforms it."*[21] With my definition of cosmography I tried to go a step further by making explicit that particular understandings of the world (and not only places/territories) are constructed through these practices. I sustained this argument based on one of the three layers of Blaser definitions of ontology. He borrowed this definition from the Actor Network Theory and considers that "*ontologies do not precede mundane practices, but rather are shaped through the practices and interactions of both humans and nonhumans.*"[22] This clearly contrasts Little's approach in which cosmographical ideas are attached to spaces to construct a place or a territory. It was using my redefined cosmography concept and focusing mainly on practices that I achieved to understand, paraphrasing Villagra[23], the changes experienced by the Angaité without denying some continuities while at the same time keeping in sight the asymmetric relations in which they were involved.

---

[1] Tim Ingold, 2000
[2] Mario Blaser and Arturo Escobar, 2016
[3] Mario Blaser, 2013: 547
[4] Arturo Escobar, 1999
[5] Mario Blaser, 2003: 37
[6] Mario Blaser, 2010: 3
[7] Burman's review of Mario Blaser, 2013
[8] Mario Blaser, 2013: 550
[9] Carolyna Tytelman, 2016: 18
[10] Rodrigo Villagra, 2010
[11] Valentina Bonifacio and Rodrigo Villagra, 2016
[12] Carolina Tytelman, 2016: 245
[13] Mario Blaser, 2010: 11
[14] Tim Ingold, 2000: 58
[15] Mario Blaser, 2003: 95
[16] Rodrigo Villagra, 2010
[17] Valentina Bonifacio, 2013a: 511
[18] Bruce Albert, 2005: 200; Alexandre Surrallés and Pedro García Hierro, 2005
[19] Yeh and Bryan 2015
[20] Carolina Tytelman, 2016: 3
[21] Ibid., 17
[22] Blaser, 2010:3
[23] Rodrigo Villagra, 2010

Appendix

Appendix 1. Table 8. Ranches used by the villagers of La Patria in 1995 and 2015 for hunting, fishing and gathering

Vi	K		Ki		C		C2		LH		SF		L		LF		M		T		LP		3Q		LT		1		6		P		PK		U		
Ranch	1995	2015	1995	2015	1995	2015	1995	2015	1995	2015	1995	2015	1995	2015	1995	2015	1995	2015	1995	2015	1995	2015	1995	2015	1995	2015	1995	2015	1995	2015	1995	2015	1995	2015	1995	2015	Tot
1.					x				x								x	o			x																1
2.		o		o	x			o														o														o	5
3.																																o					1
4.									x								x																				0
5.	x	o	x		x	o			x								x	o	x																		3
6.					x	o			x										x							o											2
7.								o												o																	2
8.					x		x	o	x		x								x	o				o													3
9.	x		x				x	o	x																												1
10.													x																								0
11.	x	o	x	o				o			x								x			o						o		o		o					7
12.							x		x		x	o							x					o													2
13.									x			o														o											2
14				o									x													o						o					3
14	x		x	o					x						x											o						o					3
15.											¿																										0

16.															¿						¿																0
17.						o									x																						1
18.															x																						0
19.															x	o					x									o						o	3
20.															¿						¿																0
Lucero												o												o		o											3
Filadelfia		o												o		o										o											4
Loma				o						o				o																							3
Casado				o																																	

Villages: Karova Guasu (KG), Karoa'í (Ki), Carpincho (C), Colonia 24 (C24), Laguna Hu (LH), San Fernández (SF), La Leona (L), Las Flores (LF), Monte Kue (MK), Tatare (T), La Paciencia (LPac), Tres Quebrachos (3Q), Laguna Teja (LT), Primero de Mayo (1M), Seis de Marzo (6M), Paraiso (P), Puente Kaigue (PK), Urundey (U). Ranches: 1. Riacho González, 2. Yakaré, 3. Tuparendá, 4. Jahapé, 5. Mayodos, 6. Mayotres/Santa Isabel, 7. Caceres Cue/Don Guido, 8. Buena Vista, 9. Esperanza, 10. Santa Cecilia, 11. Carmelitas, 12. Santa Rosa, 13. Timboty, 14. Laguna Rey, 15. Jaguar, 16. Area 16, 17. Riacho Paraguay, 18. Carpa Blanca, 19. Mirian Victoria, 20. San Pedro

Appendix 2. Celebration letter (Guarani version)

"Comunidad Indígena del Pueblo Angaité
Comunidad Indígena La Patria
Comunidad La Patria oike va'ekue pueblo angaite año 1983 entre 1986. Ojoguavae'kue misionero anglicano paraguayo.
La Patria ha'e yvy virgen, ndaipori mba'eve, ka'aguy nandi. Orekó 22.520 hectareas campo, oreko ka'aguy, ka'aguy lalera, karandayty, partillar, laguna, riacho.
Año 1983, ojeike ramo La Patria ndorekoi tapé, ndorekoi óga, ndoguerekoi Puesto de Salud, Escuela.
Pe tiempope oike 1500 personas, del pueblo angaité. Ko'anga 2015pe ofigurama 3000 censope. Misioneros Anglicano ojogua la yvy ha ome'ejey indígena Angaite imba'e voi, oje'ehaicha constitución nacional ha ley indígena 904.
Uperoguare 10 comunidad oï, ha 20 líder oï pea apytepe. Pe 20 líder ojereuní, mba'eicha oconseguita la título ha mavapa ikatú otramitá. Uperoguare año 1991, 1992, osë petei líder ojeporavo orepresenta hagua trámite de parte de título, ha pea ha'evakue Damacio Flores. Año 1992 omboguatá la documento, ha otramitá pe título. Cuatro años ojapó la trámite osehagua la título, ha mientras tanto oparticipa avei en 6 congreso nacional Anglicano, Paraguaype, Chilepe ha Argentinape. Ojapoparire la trámite ojentregá la título, año 1995pe oiko vaekue la entrega de título.
Ante la oñentregá la título, Damasio ohecha mboy persona ofirmata la título oaségurahagua herape la títulope: ha'e vaekue Ruben Fernández, Lalicio Rojas, Agüero Fernandez, Isacio Segura ha Damasio Flores. Obispo Anglicano Ellyson odesigna Damasio Flores oñangarekohagua la titulo.
Misionero otransferí la título año 1995pe, ha oentregariré oaconsejá la líderes mba'eicha oikota la ipype ha según la ley heíchaicha ndaha'ei blanco kuera mba'e. Ndikatui oike ha oiporupa rei la blanco, koa según la ley. Koa pueblo angaité mba'e ko'angagua. La Patria ha'e petei yvy comunitaria, ndaha'ei petei líder mba'e, hetä ijara, ko'anga oï 18 comunidad. Ko'anga oï 18 comunidad ha 34 líderes opytava responsable de La Patria.
Liderekuera oagradecé pastores ha Iglesia Anglicana. Rojejure avei ko'anga misioneropc ani hagua orereja, oabandoná, ore roikoteve orepytyvo hagua como anglicano. Ikaturamo romombareteve ore cultura ha religión, estudio, capacitación. Rojejure ojeinformá hagua, anglicano oïva inglaterrape ha otro tetäme. Oi heta la mitä pyahu, ha ndoikuaa mba'eicha oñemboguatá, mava la lucha ojapova'ekue, iglesia anglicano ome'ehagua, ko yvy ndaha'ei ome'erei. Ha'ekuera omboguatá la reglamento, aní otro ojapó vaí ko yvy koare, ko yvy oreko ireglamento.
Rome'e avei gracias anglicanokuera ojapouka la tupaokuera la comunidadpe. Ome'ehagua yvy la anglicanokuera heta año de lucha ogueraha de parte de andministrativa. Haupei oguahe la itiempo ome'ehagua la yvy ha hetä cosas omboja'o cuatro asentamiento, animal, vaca, cavaju, carro, tractor. Pe tiempope la ancianokuera saí oreko la experiencia, saí oprendé mba'eicha administrá, ha ko'angagua ojekambia, ko'anga la nueva generación oikua'a mba'eicha oadministrata, mba'eicha omanteneta oviviporä hagua."

Appendix 3. Celebration letter (Spanish version)

"15 de Agosto del 2015
Comunidad Indígena La Patria, del Pueblo Angaité
La comunidad La Patria fue poblada entre el año 1983 y 1986. La tierra fue comprada por los Misioneros Anglicanos del Paraguay.
La Patria era una tierra virgen, solo había monte. Tiene 22.520 hectáreas. Había campo, monte, monte lalera, palmar, espartillar, lagunas y riachos. En el año 1983 no había caminos, casas, no teníamos Puesto de Salud ni Escuelas.
En aquel tiempo entraron a La Patria 1500 personas. Ahora ya somos más de 3000 habitantes. Los Misioneros compraron la tierra y le devolvieron a los Angaité lo que era luego suyo, como lo dice la Constitución Nacional y la ley 904.
En aquel tiempo había 10 comunidades y 20 líderes. En el año 1992 los líderes se reunieron y nombraron a Damasio Flores para que gestione y represente a los Angaité y los trámites para el título. Cuatro años de gestión se tardó, y en 1995 se hizo la entrega del título. Damasio Flores recomendó que 5 personas firmen el título: Ruben Fernandez, Lalicio Rojas, Agüero Fernandez, Isacio Segura y Damasio Flores. Los Misioneros transfirieron el título en el año 1995, y aconsejaron a los líderes cómo debían vivir en ella. Como dice la ley, La Patria no es de los blancos. No pueden entrar nomás los blancos y usar. La Patria es una tierra comunitaria, no es de un líder, tiene muchos dueños, ahora ya hay 18 comunidades.
Ahora son 18 comunidades y 34 líderes responsables de La Patria.
Queremos agradecer a los Misioneros Anglicanos. Queremos pedirles que no nos abandonen, queremos que nos ayuden a fortalecer nuestra cultura y nuestra religión, con estudios y capacitación. Pedimos que informen de este mensaje a los Anglicanos que viven en Inglaterra y en otros países.
Hay muchos niños nuevos que no conocen la historia de lucha, no saben que esta tierra no fue fácil conseguir. Los Anglicanos dejaron un reglamento, para que otros no vengan a hacernos daño.
Queremos agradecer a los Anglicanos por construir las Iglesias en algunas de las comunidades.
Muchos años de gestión se tardó para asegurar esta tierra. Muchas cosas se trajo a los primeros asentamientos, animales, vacas, caballos, carros, tractor. Los ancianos no tenían experiencia como administrar eso. Ahora la nueva generación ya sabe cómo administrar, cómo mantenerse para vivir bien."

Appendix 4. Mr. Oviedo document

"15 de Agosto de 2015

Comunidad La Leona, de la Patria

Presentación de la historia de la comunidad Angaite Koyalhoc

Formación de la Junta Directiva

Los líderes de la iglesia se reunieron con los misioneros anglicanos en Makxawaya para formar la Junta Directiva en la presencia del obispo Alejandro de la Iglesia Anglicana Paraguaya y el misionero Pedro Faulkner del 14 de diciembre de 1977.

Se presentaron representantes de las personas que vivían en las estancias en Makxawaya para formar la Junta Directiva durante los festejos navideños para buscar tierra perteneciente a Puerto Pinasco. Estas tierras y zonas fueron recorridos por el obispo Alejandro, el pastor Felix Bogado, el pastor Zacarías Severo, David Rojas, Andrés Tomen, el pastor Juan Carlos, Luciano Riquelme, Chamorro Ayala, Rogelio Rondán, pastor Ricardo, Marciano Paraguari, Tomás Fernández, Agüero Fernandez, Senón Nuñez, Calí Sanabria, Pedro Alarcón, Sanchez Cerro, Alejandro Recalde y Alonso Pinasco.

Compra de la tierra

La Iglesia Anglicana compró 22.520 hectáreas en el mes de agosto del año 1982 a través del proyecto La Herencia dirigido por el misionero Pedro Tyson.

Asentamiento de la tierra

La gente de las estancias se juntó en Makxawaya y de ahí salieron para recorrer la tierra comprada para que cada grupo pudiese decidir en qué lugar de La Patria asentarse y formar sus comunidades.

El proceso de llegar allá y asentarse fue difícil por causa de la ausencia de caminos y la falta de cantinas. La gente hizo ranchos de techo capi'i y comió karanday y ysypo. No había yerba ni jabón. La gente se mantuvo a través de la caza y pesca.

Senón Nuñez fundó la comunidad de Laguna Hü y fue responsable por dicha comunidad. Agüero Fernández y Tomás Fernandez fundaron la comunidad de Corralón-Cué. José Landi y Sánchez Cerro fundaron la comunidad de Colonia Veinticuatro. Luciano Riquelme fundó la comunidad de Tatarë. Julián Rolón e Hilario González fundaron la comunidad de Carpincho. Andrés Tomen y Alejandro Recalde fundaron la comunidad de La Leona.

Pedro Alarcón y Ricardo Ramirez fundaron la comunidad de Caroa Guasu. Calí Sanabria y Rogelio Ronán fundaron la comunidad de Riacho González que ahora se llama La Paciencia. Juan Carlos Cabañas y Alón Cabañas fundaron la comunidad de Las Flores. Al principio hubo diez comunidades con un total de veinte líderes que integraron la Junta Directiva. Ellos decidieron que la tierra comprada se llamaría La Patria.

Después de la compra de la tierra la gente pudo vivir con tranquilidad.

Miembros de la Junta Directiva ya fallecidos y que ya descansan después de la lucha por la tierra

1. Agüero Fernández de Corraón-cue
2. Senón Núñez de Laguna Hü
3. Calí Sanabria de Riacho González
4. Rogelio Rondán de Riacho González
5. Ricardo Ramírez de Caroa Guasu
6. Pedro Alarcón de Caroa Guasu
7. Juan Carlos Cabañas de Las Flores
8. Alón Cabañas de Las Flores
9. Chamorro Ayala de la comunidad de Carpincho
10. Jualián Rolón de Tatarë'i
11. Hilario González de Tatarë'i
12. Andrés Tomen de La Leona

Miembros de la Junta Directiva que viven en La Patria

1. Luciano Riquelme de Tatare
2. Tomás Fernandez de Puente Kaigue
3. Marciano Paraguari de Carpincho
4. Alonso Pinasco de Carovai
5. Alenjandro Recalde de La Leona
6. Sanchez Cerro de Comunidad 24

Líderes reconocidos por el INDI del Estado Paraguayo en año 1989

La Leona: Alejandro Recalde y Lalicio Rojas
Colonia Veinticuatro: Sánchez Cerro y José Landi
Tatarë: Luciano Riquelme y Toti Riquelme
Laguna Hü: Lino Rolón y Reinando Centurión
Tatarë'i: Julián Rolón
Corralón-cué: Agüero Fernández, Tomás Fernández, Rubén Fernández
Las Flores: Alón Cabañas, Basilio Sosa
Riacho González: Calí Sanabria
Carpincho: Capata'i Segura e Isasio Segura

Caroa Guasu: Damasio Flores y Alonso Pinasco

Titulación de la tierra

Estas personas se dedicaron a los trabajos relacionados con la titulación de la tierra:

Damasio Flores, Rubén Fernández y Lino Rolón trabajaron para conseguir los títulos de propiedad de La Patria, a veces saliendo a pie de la comunidad de La Patria hasta la Ruta Trans Chaco, caminando 120 kilómetros.

Entrega de título

La ceremonia de entrega de títulos se realizó en la comunidad de Corralón-cué/San Fernández en el año 1995 con la presencia de una delegación de representantes de la Iglesia Anglicana Paraguaya encabezada por el obispo Juan Ellison, el administrador Teodosio Rivas, la abogada Filomena, los pastores Elvio Cabañas, Lucio Romero. La Comunidad de La Patria fue representada por Alejandro Recalde, Rubén Fernández, Lalicio Rojas, Luciano Riquelme, Julián Rolón, Isasio Segura, Calí Sanabria, Basilio Sosa, Alón Cabañas, Damasio Flores, Lino Rolón, Tomás Fernández, Reinaldo Centurión y Toti Riquelme.

A continuación los nombres de los líderes reconocidos que recibieron y firmaron los títulos de propiedad entregados por el obispo Juan Ellison:

Damasio Flores, Isasio Segura, Lalicio Rojas, Agüero Fernández, Rubén Fernández.

Una copia del título de propiedad fue entregada a un niño, Mauricio Fernández en representación de la Comunidad.

Dos de los firmantes originales, Isasio Segura y Agüero Fernández, ya han fallecido.

Expansión de las comunidades

Hoy en día existen 18 comunidades, que son:

Laguna Hü; La Leona; San Fernández (ex Corralón-cué); Carpincho; Tatarë; Carova Guasu; La Paciencia; Las Flores; Colonia Veinticuatro; Puente Kaigue; Montekue; Carova'i; Laguna Teja; Paraiso; Urundey; Seis de Marzo; y Tres Quebrachos.

Lista de Iglesias de La Patria

La mayoría de las Iglesias son de material, pero aún hay algunas de palma.

Iglesia de Laguna Hü (material: misionero menonita)

Iglesia de La Leona (material: anglicano)

Iglesia de Carova Guasu (material: misionero menonita)

Iglesia de La Paciencia (material: anglicano)

Iglesia de Las Flores (material: misionero menonita)

Iglesia de Carpincho
Iglesia de San Fernández (material: misionero menonita)
Iglesia de Colonia Veinticuatro (material: anglicano)
Iglesia de Carova'i
Iglesia de Montekue
Iglesia de Paraiso
Iglesia de Urunde'y
Iglesia de Laguna Teja

Escuelas de La Patria

Escuela de San Fernández, a cargo de los profesores Carlos Fernández, Ceferino Chávez y Sonia García.
Escuela de Laguna Hü, a cargo del profesor Damasio Rolón
Escuela de La Leona, a cargo del profesor Gilberto Aguilera
Escuela de Paraiso, a cargo del profesor Miciades González
Escuela de Carova'i, a cargo del profesor Jorge Armada
Escuela de Puente Kaigue, a cargo del profesor Alcides Fernandez
Escuela de Laguna Teja, a cargo del profesor Graciano Zavala
Escuela Colonia Veinticuatro a cargo del profesor Oscar Oviedo y Silverio Oviedo.
Escuela de Carpincho, a cargo de los profesores Ramón y Rosana Oviedo
Escuela de Montekue, a cargo del profesor Daniel Navarro
Escuela de Carova Guasu, a cargo del profesora Eulogia Ruiz
Escuela de Seis de Marzo, a cargo del profesor Mario Navarro
Escuela de Las Flores, a cargo de los profesores Aníbal Lorenzo y Abelino Romero

Lista de pastores de la Iglesia Anglicana de La Patria

Pasto Robert Fernández de Iglesia de San Fernández
Pastor Agripino Sánchez de la Iglesia de La Leona
Pasto Niño Agüero de la Iglesia de La Paciencia
Pastor Mario Recalde de la Iglesia de Colonia Veinticuatro
Diácono Jumencinto Torres de la Iglesi de Colonia Veinticuatro

Lista de pastores de la Iglesia Misionera Mennonita

Pastores Abelino Rolón y Anario Rolón de la Iglesia de Laguna Hü
Pastor Damiano Ávalos de la Iglesia de Carova Guasu
Ayudante Gilberto Cabañas de la Iglesia de Las Flores.

Estancias desde donde vino a la gente a La Patria

1. Estancia Santa Juanita. Encabezado por Andres Tomen (translado a La Leona), Juan Carlos (translado a Las Flores).

2. Estancia Lucero. Encabezado por Tomás Fernandez, Agüero Fernadez, Ruben Fernández (transaldo a Colonia Corralón-cue)

3. Misión Isla'i (Makxawaxa). Encabezado por Senón Núñez y Mario Recalde (translado a Laguna Hü)

4. Estancia Lapacho. Encabezado por Luciano Riquelme (translado a Tatarë), Sanchez Cerro (translado a Colonia 24)

5. Estancia Karanda. Encabezado por Julián Rolón y Hilario González (translado a Tatarë'i)

6. Estancia Carmelita. Encabezado por Damasio Flores, Alonso Pinaso y Pedro Alarcón (translado a Carova Guasu).

7. Estancia Buena Vista. Encabezado por Marciano Paraguari y Ayala Chamorro (translado a Carpincho)

8. Estancia Riacho González. Encabezado por Calí Sanabria y Rogelio Doldán, se quedaron en el mismo lugar (La Paciencia)

Comunidad Laguna Hü, Agosto año 2015"

Appendix 5. Figure 14. Participatory mapping result 1. La Patria Colony with Bing imagery 2014. Original size: A0

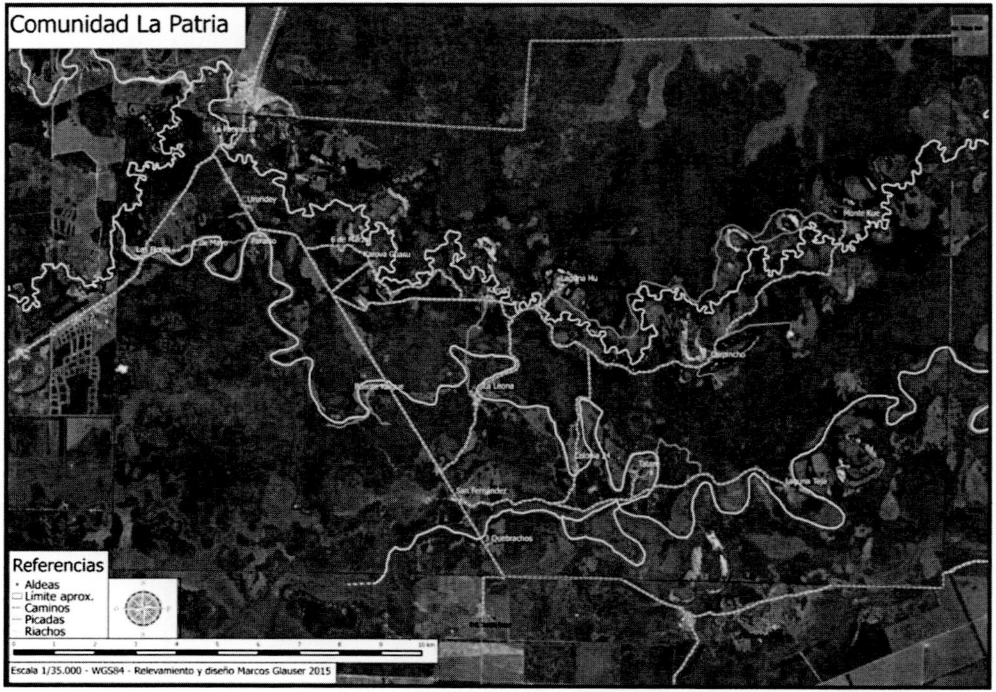

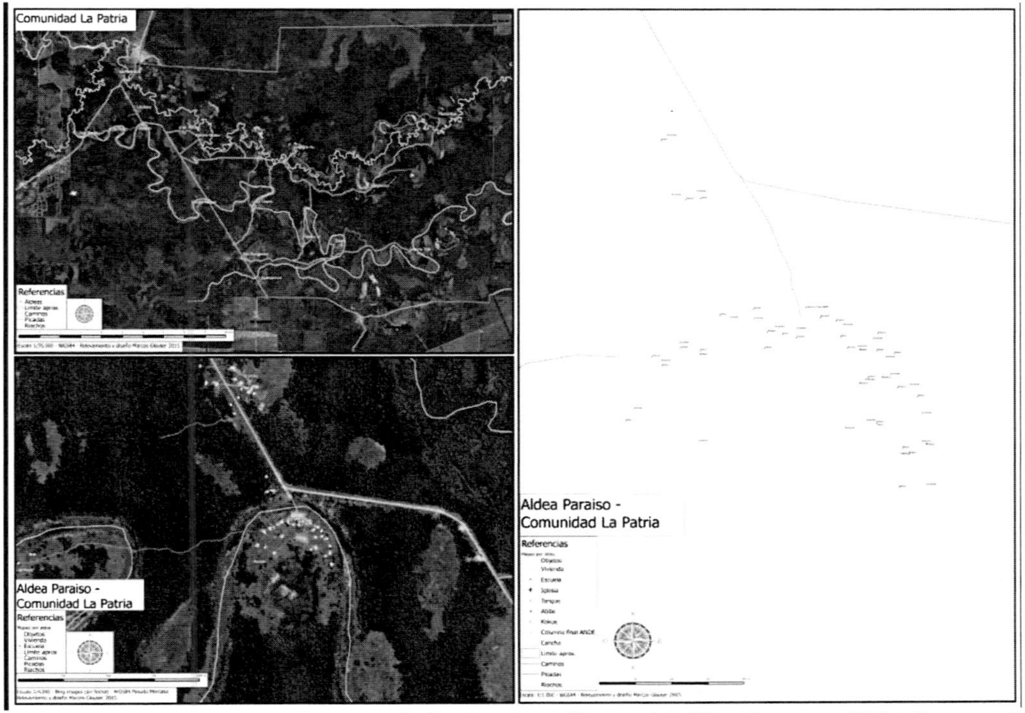

La Patria Colony with Bing imagery 2014 (top left). Paraiso village with Bing imagery 2014 (bottom left). Paraiso village (right

Appendix 6. Pictures taken by the author in La Patria.

Picture 1. Donkeys at the lagoon in Carpincho village.

Picture 2. Angaité family at their kitchen, Tatarē village.

Picture 3. Fishing with the cone trap, Las Flores village.

Picture 4. Fishing with the spear, Tatarē village

Picture 5. Celebrating youth day, San Fernández village.

Picture 6. At church, Urunde'y village.

Picture 7. Three houses, Karoa'í village.

Picture 8. Participatory mapping in Tatarē village

Picture 9. Angaité leader showing his sketch map during the transect walk.

Picture 10. Leader's assembly meeting in La Leona village.

Picture 11. Leader's assembly meeting in La Leona village.

Picture 12. Collective slaughtering of a Tapir, Tatarẽ village.

Picture 13. Natural grassland and forest

Picture 14. La Patria anniversary celebration

Picture 15. Angaité children in Colonia 24 village

Picture 16. Leaders coordinating celebration activities

Picture 17. Families from La Patria receiving the emergency aid from the SEN

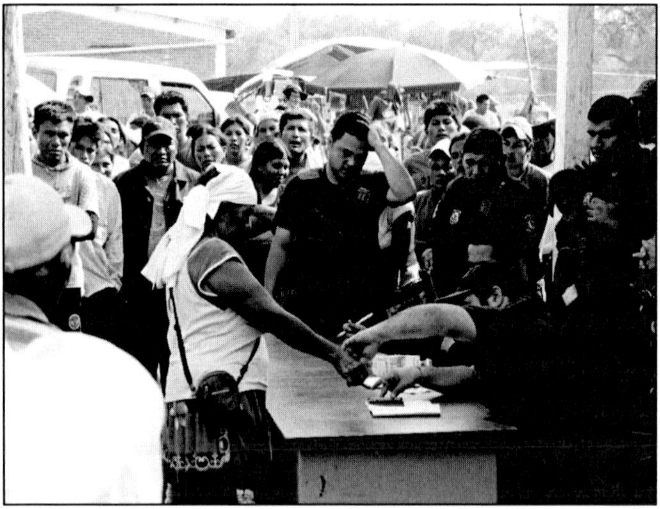

Picture 18. Angaité woman receiving the Tekoporã monetary transfer, La Leona village.

Picture 19. The author during an interview in Tatarẽ village.

Picture 20. Angaité young woman video recording the anniversary celebration.

Picture 21. Choqueo at Puente Kaigue village, April 19th 2016. Picture taken by Pascual Glauser.

Picture 22. Choqueo at Puente Kaigue village, April 19th 2016. Picture taken by Pascual Glauser

Bibliography

Albert, Bruce, 2005. Territoriality, Ethnopolitics, and Development: The Indian Movement in the Brazilian Amazon. In: Surrallés, Alexandre and García Hierro, Pedro (eds.). *The land within: Indigenous territory and the perception of the environment.* Copenhagen: IWGIA, p. 200-228

Angrosino, Michael V., 2012. *Etnografía y observación participante en investigación cualitativa.* Investigación cualitativa, vol. 3. Madrid: Morata.

Aquino, Adolfo, 2016. *Relatorio de Impacto Ambiental Estancia Enciso.* http://www.seam.gov.py/relatorio-de-impacto. Accessed 21 February 2017.

ASADEC, 2007. *Estatutos Sociales, Asociación Angaité de Desarrollo Comunitario.*

Athayde, Simone. and Marianne Schmink, 2014. "Adaptive Resistance," Conservation, and Development in the Brazilian Amazon: Contradictions of Political Organization and Empowerment in the Kaiabi Diaspora. *Ethnohistory* 61 (3): 549–574. 10.1215/00141801-2681795.

Ayala, Oscar, 2016. *Aproximación a la situación de los derechos a la tierra y el territorio de los Pueblos Indígenas en el Paraguay: Estudio sobre tierras indígenas en Paraguay.* Asunción: IEPALA

Baer, Hans A. and M. Singer, 2014. *The anthropology of climate change: An integrated critical perspective.* Routledge advances in climate change research, 1st edn. London, New York: Routledge

Barrios Mario (ed.), 2008. *Acción de Inconstitucionalidad del decreto N 11.804/08 por el cual se declara como área silvestre protegida bajo dominio privado la reserva denominada Estancia Salazar.* Asunción: TierraViva

Bartolomé, Miguel, 2000. *El Encuentro de la Gente con los Insensatos: La sedentarización de los Cazadores Ayoreo en el Chaco Paraguayo.* Asunción: CEADUC.

Bertram, Nicolas, 2016. *Cosecha soja, siembra inundación.* https://www.pagina12.com.ar/diario/sociedad/3-297922-2016-04-27.html. Accessed 21 February 2017.

Bhoola, Urmila, 2017. *End of mission statement of Urmila Bhoola, Special Rapporteur on contemporary forms of slavery, including its causes and consequences following her visit to Paraguay between 17-24 July 2017.* http://www.ohchr.org/EN/NewsEvents/Pages/DisplayNews.aspx?NewsID=21903&LangID=E. Accessed 1 June 2017.

Bird-David, Nurit, 1990. The Giving Environment: Another Perspective on the Economic System of Gatherer-Hunters. *The University of Chicado Press Journals* 31 (2): 189–196.

Blaser, Mario, 2003. *Governmentalities and authorized imaginations: A (non modern) story about indians, nature and development*. PhD thesis. McMaster University.

Blaser, Mario, 2004a. Indígenas del Chaco paraguayo: ¿Proyectos de vida o proyectos de desarrollo? *Suplemento antropológico* 39 (1): 193–229.

Blaser, Mario, 2004b. "Way of life" or "Who decides": Development, Paraguayan indigenism, and the Yshiro people's life projects. In: M. Blaser, H. A. Feit and G. McRae (eds.). *In the Way of Development: Indigenous Peoples, Life Projects and Globalization*: Zed Books.

Blaser, Mario, 2010. *Storytelling globalization from the Chaco and beyond.* New ecologies for the twenty-first century. Durham, NC: Duke University Press.

Blaser, Mario, 2013. Ontological Conflicts and the Stories of Peoples in Spite of Europe: Toward a Conversation on Political Ontology. *Current Anthropology* 54 (5).

Blaser, Mario and Arturo Escobar, 2016. Political Ecology. In: Joni Adamson, William A. Gleason and David N. Pellow (eds.). *Keywords for environmental studies*, 164-172. New York: New York University Press.

Bonifacio, Valentina, 2013a. Building up the collective: A critical assessment of the relationship between indigenous organisations and international cooperation in the Paraguayan Chaco. *Social Anthropology/Anthropologie Sociale* 21 (4): 510–522. 10.1111/1469-8676.12045.

Bonifacio, Valentina, 2013b. Meeting the Generals: A Political Ontology Analysis of the Paraguayan Maskoy Struggle for Land. *Anthropologica* 55 (2013) 385–398

Bonifacio, Valentina, 2017. Lucas Bessire, Behold the Black Caiman: A Chronicle of Ayoreo Life (Chicago, IL, and London: University of Chicago Press, 2014). *J. Lat. Am. Stud.* 49 (01): 207–209. 10.1017/S0022216X16002091.

Bonifacio, Valentina and R. Villagra, 2016. Conexiones inestables, imprevistas y pérdidas: Expandiendo la arena política en la cooperación para el desarrollo y comunidades indígenas en el Chaco paraguayo. *Revista de antropologia* 59 (3): 90–114.

Bryan, Joe, 2009. Where would we be without them?: Knowledge, space and power in indigenous politics. *Futures* 41 (1): 24–32. 10.1016/j.futures.2008.07.005.

Bryant, Raymond L. and Sinéad Bailey, 1997. *Third World political ecology*. London, New York: Routledge.

Cameron, Emile., De Leeuw, Simone and Greenwood, Margo, 2009. Indigeneity. In: N. J. Thrift and Rob Kitchin (eds.). *International encyclopedia of human geography*, pp. 352–357. Amterdam, London, Oxford: Elsevier.

Centrón, Mónica, 2016. *Relatorio de Impacto Ambiental Estancia Laguna Rey*. http://www.seam.gov.py/relatorio-de-impacto. Accessed 21 February 2017.

Chapin, Mac, Zachary Lamb and Bill Threlkeld, 2005. Mapping Indigenous Lands. *Annu. Rev. Anthropol.* 34 (1): 619–638. 10.1146/annurev.anthro.34.081804.120429.

Church Mission Society, 2015. *120 Christians confirmed in Chaco church*. https://churchmissionsociety.org/our-stories/120-christians-confirmed-chaco-church. Accessed 2 September 2017.

Combes, Isabelle, Diego Villar and Kathleen. Lowrey, 2009. Comparative Studies and the South American Gran Chaco. *Tipití: Journal of the Society for Anthropology of Lowland South America* Vol. 7.

Combès, Isabelle 2015. Prólogo. In: Córdoba, Lorena; Bossert, Federico and Richard, Nicolás (eds.). *Capitalismo en las selvas. Enclaves industriales en el Chaco y Amazonía indígena (1850-1950).*

Córdoba, Lorena; Bossert, Federico; Richard, Nicolás (eds.) 2015. *Capitalismo en las selvas: Enclaves industriales en el Chaco y Amazonía indígenas (1850-1950).* San Pedro de Atacama, Ediciones del Desierto.

Descola, Philippe, 2005. Ecology as Cosmological Analysis. In: Alexandre Surrallés and Pedro García Hierro (eds.). *The land within: Indigenous territory and the perception of the environment*. Copenhagen: IWGIA.

DGEEC, 2004. *Atlas de las Comunidades Indígenas en el Paraguay*. Fernando de la Mora, Paraguay.

DGEEC, 2005. *Poblacion Indígena*. Fernando de la Mora, Paraguay.

DGEEC, 2014. *Pueblos Indígenas en el Paraguay.: Resultados Finales de Población y Viviendas 2012*. Fernando de la Mora, Paraguay.

DGEEC, 2015. *Censo de Comunidades de los Pueblos Indígenas: Resultados finales 2012*. Fernando de la Mora, Paraguay.

Escobar, Arturo, 1995. *Encountering development: The making and unmaking of the Third World*. Princeton studies in culture/power/history. Princeton N.J.: Princeton University Press.

Escobar, Arturo, 1996. Constructing nature: Elements for a poststructural political ecology. In: Richard Peet and Michael Watts (edt) (ed.). *Liberation Ecologies: Environment, Development, Social Movements*. New York: Routledge.

Escobar, Arturo, 1999. After Nature. *Current Anthropology* 40 (1): 1–30. 10.1086/515799.

Escobar, Arturo, 2010. Postconstructivist political ecologies. In: M. R. Redclift and Graham Woodgate (eds.). *The international handbook of environmental sociology*, 2nd edn. Cheltenham, UK, Northampton, MA: Edward Elgar.

Escobar, Arturo, 2016. Thinking-feeling with the Earth: Territorial Struggles and the Ontological Dimension of the Epistemologies of the South. *AIBR* 11 (1): 11–32. 10.11156/aibr.110102e.

Esteva, Gustavo, 2010. Development. In: Wolfgang. Sachs (ed.). *The Development Dictionary: A Guide to Knowledge as Power*, pp. 1–21. New York, NY: Zed Books.

Fabre, Alain, 2016. *Los pueblos del Gran Chaco y sus lenguas, primera parte: Los enlhet-enenlhet del Chaco Paraguayo*. http://www.ling.fi/AF_Publications/SA%2040-1_Enlhet.pdf. Accessed 26 October 2016.

Ferguson, James, 1990. *The anti-politics machine: "development," depoliticization, and bureaucratic power in Lesotho / James Ferguson*. Cambridge: Cambridge University Press.

Ferguson, James, 1997. Anthropology and Its Evil Twin: Development in the Constitution of a Discipline. In: Frederick Cooper and Randall M. Packard (eds.). *International development and the social sciences: Essays on the history and politics of knowledge / edited by Frederick Cooper and Randall Packard*. Berkeley, Calif., London: University of California Press.

Fisher, William H., 2000. *Rain forest exchanges: Industry and community on an Amazonian frontier*. Smithsonian series in ethnographic inquiry. Washington, D.C.: Smithsonian Institution Press.

Galeano, Luis, 2010. *Las dinámicas en el mercado de la tierra en América Latina. El caso de Paraguay*. Asunción, Paraguay.

García Hierro, Pedro and Surrallés, Alexandre, 2009. *Antropología de un derecho: Libre determinación territorial de los pueblos*: IWGIA.

Gibbs, Graham, 2007. *Analyzing Qualitative Data*. The SAGE Qualitative Research Kit / ed. Uwe Flick, [6]. Los Angeles: SAGE Publications.

Glauser, Marcos, 2009. *Extranjerización del territorio paraguayo*. Asunción, Paraguay: BASE; RLS.

Gordillo, Gastón, 1993. La actual dinámica económica de los cazadores recolectores de Gran Chaco y los deseos imaginarios del esencialismo. *PUBLICAR-En Antropología y Ciencias Sociales* (3).

Grubb, Warbroocke, 1911. *An unknown people in an unknown land*. London.

Grupo Sunu, 2007. *Waika*. Asunción, Paraguay.

Guereña, Arantxa and Villagra, Luis, 2016. *YVY-JARA: Los dueños de la tierra en Paraguay*. Asunción, Paraguay.

Gupta, Akhil and James Ferguson, 1997. *Culture, power, place: Explorations in critical anthropology*. Durham N.C.: Duke University Press.

Guyra Paraguay, 2014. *Informe deforestación: Informe monitoreo*. http://guyra.org.py/informe-deforestacion/. Accessed 21 February 2017.

Haan, Leo de, 2012. The livelihood approach: A critical exploration. *Erdkunde* 66 (4): 345–357. 10.3112/erdkunde.2012.04.05.

Halbmayer, Ernst and Elke Mader, 2004. Anthropology of Nature: From Connecting Nature and Culture to Crossing the Great Divide. In: Barbara Plankensteiner, Stefan Khittel and Maria Six-Hohenbalken (eds.). *Contemporary issues in socio-cultural anthropology: Perspectives and research activities from Austria*. Wien: Löcker.

Hammersley, Martyn and Paul Atkinson, 2007. *Ethnography: Principles in Practice,* 3rd edn. New York: Routledge.

Hansen, M. C. et al., 2013. High-resolution global maps of 21st-century forest cover change. *Science (New York, N.Y.)* 342 (6160): 850–853. 10.1126/science.1244693.

Harrison, S., Steve Pile and N. J. Thrift, 2004. *Patterned ground: Entanglements of nature and culture*. London: Reaktion.

Hetherington, Kregg, 2009. Privatizing the private in rural Paraguay: Precarious lots and the materiality of rights. *American Ethnologist* 36 (2): 224–241. 10.1111/j.1548-1425.2009.01132.x.

Huff, Amber, 2014. Weathering the Long Wounded Year: Livelihoods, nutrition and changing political ecologies in the Mikea Forest Region, Madagascar.

ILO, 2003. *ILO Convention on Indigenous and Tribal Peoples, 1989 (No. 169): A manual*. Geneva: ILO.

INAI, 2009. *Sistema Jaguar. Relevamiento Territorial Ley 26.160/06: Manual de Procedimientos. Sistema de información Geográfica.* https://www.desarrollosocial.gob.ar/wp-content/uploads/2015/08/4.-Relevamiento-territorial-de-comunidades-indígenas.pdf.

Ingold, Tim, 2000. *The Perception of the Environment: Essays on Livelihood, Dwelling and Skill.* London: Routledge.

Jones, Owain., 2009. After nature. Entangled worlds. In: Noel Castree (ed.). *A companion to environmental geography*. Blackwell companions to geography. Chichester U.K., Malden MA: Wiley-Blackwell.

Kalish, Hannes., 2000. *Hacia el protagonismo propio: Base conceptual para el relacionamiento con comunidades indígenas*. Ya'alve-Saanga: Nengvaanemquescama Nempayvaam Enlhet.

Kidd, Stephen W., 1992. *Religious change: A case-study amongst the Enxet of the Paraguayan Chaco*. M.A Thesis.

Kidd, Stephen W., 1995. Land, Politics and Benevolent Shamanism: The Enxet Indians in a Democratic Paraguay. *J. Lat. Am. Stud.* 27 (01): 43. 10.1017/S0022216X00010166.

Kidd, Stephen W., 2000a. Knowledge and the practice of love and hate among the Enxet of Paraguay. In: J. Overing and A. Passes (eds.). *The Anthropology of Love and Anger. The aesthetics of conviviality in Native Amazonia.*

Kidd, Stephen W., 2000b. *Love and hate among the people without things: The social and economic relations of the Enxet people of Paraguay.*

Kleinpenning, J. M. G., 2009. *Rural Paraguay 1870-1963: A geography of progress, plunder and poverty*. Bibliotheca Ibero-americana, v. 127. Madrid, Frankfurt am Main: Iberoamericana; Vervuert.

Krebs, Edgardo and José Braunstein, 2001. The Renewal of Gran Chaco Studies. *History of Anthropology Newsletter* Vol. 28 (1).

Lampert, Alfredo, 2015. *Relatorio de impacto ambiental Estancia 14 de Mayo.* http://www.seam.gov.py/relatorio-de-impacto. Accessed 21 February 2017.

Latour, Bruno, 2005. *Reassembling the social: An introduction to actor-network-theory / Bruno Latour.* Clarendon lectures in management studies. Oxford: Oxford University Press.

Leake, Andrew, 1998. *Subsistence and land-use amongst resettled indigenous people in the paraguayan Chaco: A participatory approach*. PhD.

Leake, Andrew, 2006. *A study of the extent to which the land of Santa Elisa and Michí (14,404 hectares) will be sufficient in size and quality to enable the indigenous people of Sawhoyamaxa (Paraguayan Chaco) to preserve and develop their particular way of life.: A study requested by the Inter-American Court of Human Rights, in relaiton to Case: 12.419 Sawhoyamaxa-Paraguay.*

Little, Paul E., 2001. *Amazonia: Territorial struggles on perennial frontiers.* Center books in natural history. Baltimore: Johns Hopkins University Press.

MAG, 2008. *Censo Agropecuario Nacional.* Asunción, Paraguay.

MAG, 2014. *Síntesis Estadísticas. Producción Agropecuaria año agrícola 2013/2014.*

Marschke, Melissa., and Fikret. Berkes, 2006. Exploring strategies that build livelihood resilience: a case from Cambodia. *Ecology and Society.*

Martínez Alier, Joan, 2004. Political ecology, distributional conflicts, and economic incommensurability. *Sustainability: Sustainable development* 3.

Massey, Doreen, 2005. Negotiating Nonhuman/Human Place. *Antipode* 37 (2): 353–357. 10.1111/j.0066-4812.2005.00497.x.

Meliá, Bartomeu and Adriana Almada, 2011. *Mundo guaraní,* 2nd edn. Asunción, Paraguay: Banco Interamericano de Desarrollo; Servilibro.

Métraux, Alfred, 1946. Ethnography of the Chaco. In: Julian Steward (ed.). *Handbook of South American Indians.* Washington: U.S G.P.O.

Neumann, Roderick P., 2005. *Making political ecology.* Human geography in the making. London, New York: Hodder Arnold; Distributed in the United States of America by Oxford University Press.

OAS, 2010. *Indigenous and tribal people's rights over their ancestral lands and natural resources: Norms and jurisprudence of the Inter-American human rights system.*

Obrist, Brigit., Pfeiffer, Constanze and Henley, Robert, 2010. Multi-layered social resilience: A new approach in mitigation research. *Progress in Development Studies* 10 (4): 283–293. 10.1177/146499340901000402.

Overing, Joanna and Passes, Alan, 2000. Introduction Conviviality and the opening up of Amazonian anthropology. In: J. Overing and A. Passes (eds.). *The Anthropology of Love and Anger. The aesthetics of conviviality in Native Amazonia.*

Pastore, Carlos, 2008. *La lucha por la tierra en el Paraguay,* 3rd edn. Asunción, Paraguay: Intercontinental Editora.

Perreault Tom, Gavin Bridge and James McCarthy (eds.), 2015. *The Routledge Handbook of Political Ecology*: Routledge.

PNC, 2012. *Paraguayan National Constitution.*

Rapport, Nigel and J. Overing, 2003. *Social and Cultural Anthropology: The Key Concepts*: Routledge.

Renshaw, John, 2002. *The Indians of the Paraguayan Chaco: Identity and economy*. Lincoln: University of Nebraska Press.

Rodriguez, Carolina and Marcos Glauser, 2014. Mapeo Participativo en parte del Tekoha Guasú, Territorio Mbyá Guaraní. *Ava Revista de Antropología* 24.

Rubenstein, Steven L., 2004. Steps to a Political Ecology of Amazonia. *Tipití: Journal of the Society for Anthropology of Lowland South America* 2 (2): 131–175.

Sachs Wolfgang. (ed.), 2010. *The Development Dictionary: A Guide to Knowledge as Power*. New York, NY: Zed Books.

Sakdapolrak, P., 2014. Livelihoods as social practices – re-energising livelihoods research with Bourdieu's theory of practice. *Geogr. Helv.* 69 (1): 19–28. 10.5194/gh-69-19-2014.

Salamanca Carlos and Rosario Espina (eds.), 2012. *Mapas y derechos: Experiencias y aprendizajes en América Latina*. Rosario Argentina: UNR Editora.

Scoones, Ian, 2009. Livelihoods perspectives and rural development. *The Journal of Peasant Studies* 36 (1): 171–196. 10.1080/03066150902820503.

Strathern, Marilyn, 1996. Cutting the Network. *The Journal of the Royal Anthropological Institute* 2 (3): 517–535.

Surrallés Alexandre and Pedro García Hierro (eds.), 2005. *The land within: Indigenous territory and the perception of the environment*. Copenhagen: IWGIA.

Tauli-Corpuz, Victoria, 2015. *Report of the Special Rapporteur on the rights of indigenous peoples*.

Thrift, N. J., 1999. Steps to an ecology of place. In: Doreen B. Massey, John Allen and Philip Sarre (eds.). *Human geography today*. Malden, Mass.: Polity Press.

Tola, Florencia C., Celeste Medrano and Lorena Cardin, 2013. *Gran Chaco: Ontologías, poder, afectividad*. Ethnographica. Buenos Aires: Asociación Civil Rumbo Sur.

Tytelman, Carolina, 2016. *Place and Forest Co-management in Nitassinan/Labrador*. Doctor of Philosophy. Newfoundland.

United Nations, 2008. *United Nations declaration on the rights of indigenous peoples: Adopted by the General Assembly on 13 September 2007*. Geneva: OHCHR.

Unruh, Ronaldo, 2015. *Nengvaanemkeskama Nempayvaam Enlhet*. http://www.enlhet.org/analisis.html. Accessed 31 May 2016.

Villagra, Rodrigo, 2010. *The two shamans and the owner of the cattle: alterity, stroytelling and shamanism amongs the Angaite of the paraguayan Chaco*. PhD Thesis.

Villagra, Rodrigo, 2014. *Maike makha valayo. No habían paraguayos: Reflexiones etnográficas en torno a los angaité del Chaco*. Asunción, Paraguay: CEADUC.

VMME, 2007. *Identificación de necesidades y acciones en la comunidad beneficiaria en Paraguay: Proyecto Algarrobo*.

Von Bremen, Volker, 1987. *Fuentes de caza y recolección modernas: Proyectos de ayuda al desarrollo destinados a los indígenas del Gran Chaco Argentina, Paraguay, Bolivia*.

Von Bremen, Volker, 1994. La significación del derecho a la tenencia de la tierra para los pueblos tradicionalmente no-sedentarios del chaco paraguayo. *Suplemento antropológico* XXIX (1-2).

Von Bremen, Volker, 2017. Indigenous Deals: Cosmologies Negotiated in Environmental and Development Projects. *Sociologus* 67: 43–57.

Wolf, Eric, 1972. Ownership and Political Ecology. *Anthropological Quarterly* 45 (3): 201–205. 10.2307/3316532.

Yanosky, Alberto., 2017. *Monitoreo Mensual del Cambio de Uso de Cobertura de la Tierra, Incendios y Variación de la Cubirta de Aguas en el Gran Chaco Americano*.

Yeh, Emily and Joe Bryan, 2015. Indigeneity. In: Tom Perreault, Gavin Bridge and James McCarthy (eds.). *The Routledge Handbook of Political Ecology*, p. 531: Routledge.

List of acronyms, figures, tables and boxes

Acronyms

APICSA - Apicultura del Chaco Sociedad Anónima
ASADEC - Asociación Angaité de Desarrollo Comunitario
ASCIM - Asociación de Servicios de Cooperación Indígena Mennonita
CLIBCH - Coordinadora de Líderes del Bajo Chaco
IAHRS - Inter-American Human Rights System
INDI - Instituto Paraguayo del Indígena
ILO - International Labour Organization
IPC - International Products Corporation
PRODECHACO - Programa de Desarrollo Sustentable del Chaco Paraguayo
RIMA - Relatorio de Impacto Ambiental
SAMS - South American Missionary Society
SEAM - Secretaría del Ambiente
SEN - Secretaría de Emergencia Nacional
UNDP – United Nations Development Programme

Figures

Tables

Boxes

About the author

Marcos Glauser (35) is a researcher at PRONII CONACYT-Paraguay and holds a PhD in Sociocultural Anthropology from the Philipps-Universität Marburg, Germany. Since 2008 he conducts research and has published articles and books on food and territorial management of indigenous and peasant communities as well as social and environmental impacts of mining, land grabbing and the large-scale soy production in Paraguay. He is a board member of three Paraguayan NGOs that ensure the human rights and self-determination of indigenous peoples.

Action Anthropology/Aktionsethnologie

Thomas Niederberger; Tobias Haller; Helen Gambon; Madlen Kobi; Irina Wenk (Eds.)
The Open Cut
Mining, Transnational Corporations and Local Populations
The rapid expansion of the mineral and metal mining sector in the past decade was accompanied by an increase in social conflicts. What are the impacts of large-scale mining operations? What are the strategies used by transnational corporations to gain access to underground resources and legitimize their activities? And how do local and indigenous communities confronted with mining react to, negotiate with and resist these activities? This book covers 13 case studies of copper, gold, uranium and other mining operations, situated in Latin America, Africa, Asia, Australia and Switzerland. With an extensive introduction to the subject and a systematic comparison across mining operations in different phases of development and social contexts, it serves as a primer and reference book for activists, students and researchers alike.
vol. 2, 2016, 538 pp., 49,90 €, pb., ISBN 978-3-643-80151-7

Ethnologie / Anthropology

Manuela Schmundt Wyss
Jaku'eke! – Auf, lasst uns uns bewegen!
Organisation als Praxis am Beispiel der Organización Campesina del Norte, Nordostparaguay
Menschen gestalten Organisationen, als Praxis, aber auch als Struktur und Ergebnis von Praxen. Dieses Buch bietet Einblick in alltägliche Organisationspraxen der Mitglieder einer regionalen paraguayischen Bauernorganisation. Die Praxen sind Teil eines umfassenden Repertoires von Handlungspraxen in Zeiten sich stark verändernder Lebens- und Deutungszusammenhänge mit lokalen, regionalen und nationalen sowie transnationalen Bezügen. Im Zentrum all dieser Bezüge stehen in der vorliegenden Ethnografie die Akteurinnen und Akteure, deren grundlegendes Ziel die Gestaltung würdiger und tragfähiger Lebensumstände für die organisierten campesinas/campesinos und ihre Familien ist.
Bd. 45, 2012, 504 S., 50,90 €, br., ISBN 978-3-643-80104-3